A Brief History of Mathematics

KARL FINK

WOOSTER BEMAN AND DAVID SMITH, TRANSLATORS

NEW YORK

A Brief History of Mathematics

For information, address:

Cosimo, P.O. Box 416
Old Chelsea Station
New York, NY 10113-0416

or visit our website at:
www.cosimobooks.com

A Brief History of Mathematics was originally published in 1900.

Cover design by www.kerndesign.net

ISBN: 978-1-60206-385-3

EDITORS' PREFACE.

The volume called Higher Mathematics, the first edition of which was published in 1896, contained eleven chapters by eleven authors, each chapter being independent of the others, but all supposing the reader to have at least a mathematical training equivalent to that given in classical and engineering colleges. The publication of that volume is now discontinued and the chapters are issued in separate form. In these reissues it will generally be found that the monographs are enlarged by additional articles or appendices which either amplify the former presentation or record recent advances. This plan of publication has been arranged in order to meet the demand of teachers and the convenience of classes, but it is also thought that it may prove advantageous to readers in special lines of mathematical literature.

It is the intention of the publishers and editors to add other monographs to the series from time to time, if the call for the same seems to warrant it. Among the topics which are under consideration are those of elliptic functions, the theory of numbers, the group theory, the calculus of variations, and non-Euclidean geometry; possibly also monographs on branches of astronomy, mechanics, and mathematical physics may be included. It is the hope of the editors that this form of publication may tend to promote mathematical study and research over a wider field than that which the former volume has occupied.

December, 1905.

AUTHOR'S PREFACE.

THIS little work was published about ten years ago as a chapter in Merriman and Woodward's Higher Mathematics. It was written before the numerous surveys of the development of science in the past hundred years, which appeared at the close of the nineteenth century, and it therefore had more reason for being then than now, save as it can now call attention to these later contributions. The conditions under which it was published limited it to such a small compass that it could do no more than present a list of the most prominent names in connection with a few important topics. Since it is necessary to use the same plates in this edition, simply adding a few new pages, the body of the work remains substantially as it first appeared. The book therefore makes no claim to being history, but stands simply as an outline of the prominent movements in mathematics, presenting a few of the leading names, and calling attention to some of the bibliography of the subject.

It need hardly be said that the field of mathematics is now so extensive that no one can longer pretend to cover it, least of all the specialist in any one department. Furthermore it takes a century or more to weigh men and their discoveries, thus making the judgment of contemporaries often quite worthless. In spite of these facts, however, it is hoped that these pages will serve a good purpose by offering a point of departure to students desiring to investigate the movements of the past hundred years. The bibliography in the foot-notes and in Articles 19 and 20 will serve at least to open the door, and this in itself is a sufficient excuse for a work of this nature.

TEACHERS COLLEGE, COLUMBIA UNIVERSITY,
December, 1905.

CONTENTS.

Art.			Page
	1.	Introduction	7
	2.	Theory of Numbers	11
	3.	Irrational and Transcendent Numbers	13
	4.	Complex Numbers	15
	5.	Quaternions and Ausdehnungslehre	17
	6.	Theory of Equations	19
	7.	Substitutions and Groups	24
	8.	Determinants	26
	9.	Quantics	28
	10.	Calculus	31
	11.	Differential Equations	35
	12.	Infinite Series	39
	13.	Theory of Functions	43
	14.	Probabilities and Least Squares	50
	15.	Analytic Geometry	52
	16.	Modern Geometry	58
	17.	Trigonometry and Elementary Geometry	63
	18.	Non-Euclidean Geometry	65
	19.	Bibliography	68
	20.	General Tendencies	74
		Index	79

HISTORY OF MODERN MATHEMATICS.

ART. 1. INTRODUCTION.

In considering the history of modern mathematics two questions at once arise: (1) what limitations shall be placed upon the term Mathematics; (2) what force shall be assigned to the word Modern? In other words, how shall Modern Mathematics be defined?

In these pages the term Mathematics will be limited to the domain of pure science. Questions of the applications of the various branches will be considered only incidentally. Such great contributions as those of Newton in the realm of mathematical physics, of Laplace in celestial mechanics, of Lagrange and Cauchy in the wave theory, and of Poisson, Fourier, and Bessel in the theory of heat, belong rather to the field of applications.

In particular, in the domain of numbers reference will be made to certain of the contributions to the general theory, to the men who have placed the study of irrational and transcendent numbers upon a scientific foundation, and to those who have developed the modern theory of complex numbers and its elaboration in the field of quaternions and Ausdehnungslehre. In the theory of equations the names of some of the leading investigators will be mentioned, together with a brief statement of the results which they secured. The impossibility of solving the quintic will lead to a consideration of the names of the founders of the group theory and of the doctrine of determinants. This phase of higher algebra will be followed by the theory of forms, or quantics. The later development of the calculus, leading to differential equations and the theory of functions, will complete the algebraic side, save for a brief reference to the theory of probabilities. In the domain of geometry some of the contributors to the later development of the analytic and synthetic fields will be mentioned, together with the most noteworthy results of their labors. Had the author's space not been so strictly limited he would have given lists of those who have worked in other important lines, but the topics considered have been thought to have the best right to prominent place under any reasonable definition of Mathematics.

Modern Mathematics is a term by no means well defined. Algebra cannot be called modern, and yet the theory of equations has received some of its most important additions during the nineteenth century, while the theory of forms is a recent creation. Similarly with elementary geometry; the labors of Lobachevsky and Bolyai during the second quarter of the century threw a new light upon the whole subject, and more recently the study of the triangle has added another chapter to the theory. Thus the history of modern mathematics must also be the modern history of ancient branches, while subjects which seem the product of late generations have root in other centuries than the present.

How unsatisfactory must be so brief a sketch may be inferred from a glance at the Index du Répertoire Bibliographique des Sciences Mathématiques (Paris, 1893), whose seventy-one pages contain the mere enumeration of subjects in large part modern, or from a consideration of the twenty-six volumes of the Jahrbuch über die Fortschritte der Mathematik, which now devotes over a thousand pages a year to a record of the progress of the science.*

The seventeenth and eighteenth centuries laid the founda-

* The foot-notes give only a few of the authorities which might easily be cited. They are thought to include those from which considerable extracts have been made, the necessary condensation of these extracts making any other form of acknowledgment impossible.

tions of much of the subject as known to-day. The discovery of the analytic geometry by Descartes, the contributions to the theory of numbers by Fermat, to algebra by Harriot, to geometry and to mathematical physics by Pascal, and the discovery of the differential calculus by Newton and Leibniz, all contributed to make the seventeenth century memorable. The eighteenth century was naturally one of great activity. Euler and the Bernoulli family in Switzerland, d'Alembert, Lagrange, and Laplace in Paris, and Lambert in Germany, popularized Newton's great discovery, and extended both its theory and its applications. Accompanying this activity, however, was a too implicit faith in the calculus and in the inherited principles of mathematics, which left the foundations insecure and necessitated their strengthening by the succeeding generation.

The nineteenth century has been a period of intense study of first principles, of the recognition of necessary limitations of various branches, of a great spread of mathematical knowledge, and of the opening of extensive fields for applied mathematics. Especially influential has been the establishment of scientific schools and journals and university chairs. The great renaissance of geometry is not a little due to the foundation of the École Polytechnique in Paris (1794–5), and the similar schools in Prague (1806), Vienna (1815), Berlin (1820), Karlsruhe (1825), and numerous other cities. About the middle of the century these schools began to exert a still a greater influence through the custom of calling to them mathematicians of high repute, thus making Zürich, Karlsruhe, Munich, Dresden, and other cities well known as mathematical centers.

In 1796 appeared the first number of the Journal de l'École Polytechnique. Crelle's Journal für die reine und angewandte Mathematik appeared in 1826, and ten years later Liouville began the publication of the Journal de Mathématiques pures et appliquées, which has been continued by Resal and Jordan. The Cambridge Mathematical Journal was established in 1839, and merged into the Cambridge and Dublin Mathematical

Journal in 1846. Of the other periodicals which have contributed to the spread of mathematical knowledge, only a few can be mentioned: the Nouvelles Annales de Mathématiques (1842), Grunert's Archiv der Mathematik (1843), Tortolini's Annali di Scienze Matematiche e Fisiche (1850), Schlömilch's Zeitschrift für Mathematik und Physik (1856), the Quarterly Journal of Mathematics (1857), Battaglini's Giornale di Matematiche (1863), the Mathematische Annalen (1869), the Bulletin des Sciences Mathématiques (1870), the American Journal of Mathematics (1878), the Acta Mathematica (1882), and the Annals of Mathematics (1884).* To this list should be added a recent venture, unique in its aims, namely, L'Intermédiaire des Mathématiciens (1894), and two annual publications of great value, the Jahrbuch already mentioned (1868), and the Jahresbericht der deutschen Mathematiker-Vereinigung (1892).

To the influence of the schools and the journals must be added that of the various learned societies † whose published proceedings are widely known, together with the increasing liberality of such societies in the preparation of complete works of a monumental character.

The study of first principles, already mentioned, was a natural consequence of the reckless application of the new calculus and the Cartesian geometry during the eighteenth century. This development is seen in theorems relating to infinite series, in the fundamental principles of number, rational,

* For a list of current mathematical journals see the Jahrbuch über die Fortschritte der Mathematik. A small but convenient list of standard periodicals is given in Carr's Synopsis of Pure Mathematics, p. 843; Mackay, J. S., Notice sur le journalisme mathématique en Angleterre, Association française pour l'Avancement des Sciences, 1893, II, 303; Cajori, F., Teaching and History of Mathematics in the United States, pp. 94, 277; Hart, D. S., History of American Mathematical Periodicals, The Analyst, Vol. II, p. 131.

† For a list of such societies consult any recent number of the Philosophical Transactions of Royal Society of London. Dyck, W., Einleitung zu dem für den mathematischen Teil der deutschen Universitätsausstellung ausgegebenen Specialkatalog, Mathematical Papers Chicago Congress (New York, 1896). p. 41.

irrational, and complex, and in the concepts of limit, continuity, function, the infinite, and the infinitesimal. But the nineteenth century has done more than this. It has created new and extensive branches of an importance which promises much for pure and applied mathematics. Foremost among these branches stands the theory of functions founded by Cauchy, Riemann, and Weierstrass, followed by the descriptive and projective geometries, and the theories of groups, of forms, and of determinants.

The nineteenth century has naturally been one of specialization. At its opening one might have hoped to fairly compass the mathematical, physical, and astronomical sciences, as did Lagrange, Laplace, and Gauss. But the advent of the new generation, with Monge and Carnot, Poncelet and Steiner, Galois, Abel, and Jacobi, tended to split mathematics into branches between which the relations were long to remain obscure. In this respect recent years have seen a reaction, the unifying tendency again becoming prominent through the theories of functions and groups.*

ART. 2. THEORY OF NUMBERS.

The Theory of Numbers,† a favorite study among the Greeks, had its renaissance in the sixteenth and seventeenth centuries in the labors of Viète, Bachet de Méziriac, and especially Fermat. In the eighteenth century Euler and Lagrange contributed to the theory, and at its close the subject began to take scientific form through the great labors of Legendre (1798), and Gauss (1801). With the latter's Disquisitiones Arithmeticæ (1801) may be said to begin the modern theory of numbers. This theory separates into two branches, the one dealing with integers, and concerning itself especially

* Klein, F., The Present State of Mathematics, Mathematical Papers of Chicago Congress (New York, 1896), p. 133.

† Cantor, M., Geschichte der Mathematik, Vol. III, p. 94; Smith, H. J. S., Report on the theory of numbers; Collected Papers, Vol. I; Stolz, O., Grössen und Zahlen, Leipzig, 1891.

with (1) the study of primes, of congruences, and of residues, and in particular with the law of reciprocity, and (2) the theory of forms, and the other dealing with complex numbers.

The Theory of Primes* has attracted many investigators during the nineteenth century, but the results have been detailed rather than general. Tchébichef (1850) was the first to reach any valuable conclusions in the way of ascertaining the number of primes between two given limits. Riemann (1859) also gave a well-known formula for the limit of the number of primes not exceeding a given number.

The Theory of Congruences may be said to start with Gauss's Disquisitiones. He introduced the symbolism $a \equiv b$ (mod c), and explored most of the field. Tchébichef published in 1847 a work in Russian upon the subject, and in France Serret has done much to make the theory known.

Besides summarizing the labors of his predecessors in the theory of numbers, and adding many original and noteworthy contributions, to Legendre may be assigned the fundamental theorem which bears his name, the Law of Reciprocity of Quadratic Residues. This law, discovered by induction and enunciated by Euler, was first proved by Legendre in his Théorie des Nombres (1798) for special cases. Independently of Euler and Legendre, Gauss discovered the law about 1795, and was the first to give a general proof. To the subject have also contributed Cauchy, perhaps the most versatile of French mathematicians of the century; Dirichlet, whose Vorlesungen über Zahlentheorie, edited by Dedekind, is a classic; Jaçobi, who introduced the generalized symbol which bears his name; Liouville, Zeller, Eisenstein, Kummer, and Kronecker. The theory has been extended to include cubic and biquadratic reciprocity, notably by Gauss, by Jacobi, who first proved the law of cubic reciprocity, and by Kummer.

* Brocard, H., Sur la fréquence et la totalité des nombres premiers; Nouvelle Correspondence de Mathématiques, Vols. V and VI; gives recent history to 1879.

To Gauss is also due the representation of numbers by binary quadratic forms. Cauchy, Poinsot (1845), Lebesque (1859, 1868), and notably Hermite have added to the subject. In the theory of ternary forms Eisenstein has been a leader, and to him and H. J. S. Smith is also due a noteworthy advance in the theory of forms in general. Smith gave a complete classification of ternary quadratic forms, and extended Gauss's researches concerning real quadratic forms to complex forms. The investigations concerning the representation of numbers by the sum of 4, 5, 6, 7, 8 squares were advanced by Eisenstein and the theory was completed by Smith.

In Germany, Dirichlet was one of the most zealous workers in the theory of numbers, and was the first to lecture upon the subject in a German university. Among his contributions is the extension of Fermat's theorem on $x^n + y^n = z^n$, which Euler and Legendre had proved for $n = 3, 4$, Dirichlet showing that $x^5 + y^5 \neq az^5$. Among the later French writers are Borel; Poincaré, whose memoirs are numerous and valuable; Tannery, and Stieltjes. Among the leading contributors in Germany are Kronecker, Kummer, Schering, Bachmann, and Dedekind. In Austria Stolz's Vorlesungen über allgemeine Arithmetik (1885–86), and in England Mathews' Theory of Numbers (Part I, 1892) are among the most scholarly of general works. Genocchi, Sylvester, and J. W. L. Glaisher have also added to the theory.

ART. 3. IRRATIONAL AND TRANSCENDENT NUMBERS.

The sixteenth century saw the final acceptance of negative numbers, integral and fractional. The seventeenth century saw decimal fractions with the modern notation quite generally used by mathematicians. The next hundred years saw the imaginary become a powerful tool in the hands of De Moivre, and especially of Euler. For the nineteenth century it remained to complete the theory of complex numbers, to separate irrationals into algebraic and transcendent, to prove the existence of transcendent numbers, and to make a scientific study of a subject which had remained almost dormant since Euclid, the theory of irrationals. The year 1872 saw the publication of the theories of Weierstrass (by his pupil Kossak), Heine (Crelle, 74), G. Cantor (Annalen, 5), and Dedekind. Méray had taken in 1869 the same point of departure as Heine, but the theory is generally referred to the year 1872. Weierstrass's method has been completely set forth by Pincherle (1880), and Dedekind's has received additional prominence through the author's later work (1888) and the recent indorsement by Tannery (1894). Weierstrass, Cantor, and Heine base their theories on infinite series, while Dedekind founds his on the idea of a cut (Schnitt) in the system of real numbers, separating all rational numbers into two groups having certain characteristic properties. The subject has received later contributions at the hands of Weierstrass, Kronecker (Crelle, 101), and Méray.

Continued Fractions, closely related to irrational numbers (and due to Cataldi, 1613),* received attention at the hands of Euler, and at the opening of the nineteenth century were brought into prominence through the writings of Lagrange. Other noteworthy contributions have been made by Druckenmüller (1837), Kunze (1857), Lemke (1870), and Günther (1872). Ramus (1855) first connected the subject with determinants, resulting, with the subsequent contributions of Heine, Möbius, and Günther, in the theory of Kettenbruchdeterminanten. Dirichlet also added to the general theory, as have numerous contributors to the applications of the subject.

Transcendent Numbers † were first distinguished from algebraic irrationals by Kronecker. Lambert proved (1761) that π cannot be rational, and that e^n (n being a rational number) is irrational, a proof, however, which left much to be desired.

* But see Favaro, A., Notizie storiche sulle frazioni continue dal secolo decimoterzo al decimosettimo, Boncompagni's Bulletino, Vol. VII, 1874, pp. 451, 533.

† Klein, F., Vorträge über ausgewählte Fragen der Elementargeometrie, 1895, p 38; Bachmann, P., Vorlesungen über die Natur der Irrationalzahlen, 1892.

Legendre (1794) completed Lambert's proof, and showed that π is not the square root of a rational number. Liouville (1840) showed that neither e nor e^2 can be a root of an integral quadratic equation. But the existence of transcendent numbers was first established by Liouville (1844, 1851), the proof being subsequently displaced by G. Cantor's (1873). Hermite (1873) first proved e transcendent, and Lindemann (1882), starting from Hermite's conclusions, showed the same for π. Lindemann's proof was much simplified by Weierstrass (1885), still further by Hilbert (1893), and has finally been made elementary by Hurwitz and Gordan.

ART. 4. COMPLEX NUMBERS.

The Theory of Complex Numbers * may be said to have attracted attention as early as the sixteenth century in the recognition, by the Italian algebraists, of imaginary or impossible roots. In the seventeenth century Descartes distinguished between real and imaginary roots, and the eighteenth saw the labors of De Moivre and Euler. To De Moivre is due (1730) the well-known formula which bears his name, $(\cos\phi + i\sin\phi)^n = \cos n\phi + i\sin n\phi$, and to Euler (1748) the formula $\cos\phi + i\sin\phi = e^{\phi i}$.

The geometric notion of complex quantity now arose, and as a result the theory of complex numbers received a notable expansion. The idea of the graphic representation of complex numbers had appeared, however, as early as 1685, in Wallis's De Algebra tractatus. In the eighteenth century Kühn (1750) and Wessel (about 1795) made decided advances towards the present theory. Wessel's memoir appeared in the Proceedings of the Copenhagen Academy for 1799, and is exceedingly

* Riecke, F., Die Rechnung mit Richtungszahlen, 1856, p. 161; Hankel, H., Theorie der komplexen Zahlensysteme, Leipzig, 1867; Holzmüller, G., Theorie der isogonalen Verwandtschaften, 1882, p. 21; Macfarlane, A., The Imaginary of Algebra, Proceedings of American Association 1892, p. 33; Baltzer, R., Einführung der komplexen Zahlen, Crelle, 1882; Stolz, O., Vorlesungen über allgemeine Arithmetik, 2. Theil, Leipzig, 1886.

clear and complete, even in comparison with modern works. He also considers the sphere, and gives a quaternion theory from which he develops a complete spherical trigonometry. In 1804 the Abbé Buée independently came upon the same idea which Wallis had suggested, that $\pm\sqrt{-1}$ should represent a unit line, and its negative, perpendicular to the real axis. Buée's paper was not published until 1806, in which year Argand also issued a pamphlet on the same subject. It is to Argand's essay that the scientific foundation for the graphic representation of complex numbers is now generally referred. Nevertheless, in 1831 Gauss found the theory quite unknown, and in 1832 published his chief memoir on the subject, thus bringing it prominently before the mathematical world. Mention should also be made of an excellent little treatise by Mourey (1828), in which the foundations for the theory of directional numbers are scientifically laid. The general acceptance of the theory is not a little due to the labors of Cauchy and Abel, and especially the latter, who was the first to boldly use complex numbers with a success that is well known.

The common terms used in the theory are chiefly due to the founders. Argand called $\cos\phi + i\sin\phi$ the "direction factor", and $r = \sqrt{a^2 + b^2}$ the "modulus"; Cauchy (1828) called $\cos\phi + i\sin\phi$ the "reduced form" (l'expression réduite); Gauss used i for $\sqrt{-1}$, introduced the term "complex number" for $a + bi$, and called $a^2 + b^2$ the "norm." The expression "direction coefficient", often used for $\cos\phi + i\sin\phi$, is due to Hankel (1867), and "absolute value," for "modulus," is due to Weierstrass.

Following Cauchy and Gauss have come a number of contributors of high rank, of whom the following may be especially mentioned: Kummer (1844), Kronecker (1845), Scheffler (1845, 1851, 1880), Bellavitis (1835, 1852), Peacock (1845), and De Morgan (1849). Möbius must also be mentioned for his numerous memoirs on the geometric applications of complex numbers, and Dirichlet for the expansion of the theory to include primes, congruences, reciprocity, etc., as in the case of real numbers.

Other types* have been studied, besides the familiar $a + bi$, in which i is the root of $x^2 + 1 = 0$. Thus Eisenstein has studied the type $a + bj$, j being a complex root of $x^3 - 1 = 0$. Similarly, complex types have been derived from $x^k - 1 = 0$ (k prime). This generalization is largely due to Kummer, to whom is also due the theory of Ideal Numbers,† which has recently been simplified by Klein (1893) from the point of view of geometry. A further complex theory is due to Galois, the basis being the imaginary roots of an irreducible congruence, $F(x) \equiv 0$ (mod p, a prime). The late writers (from 1884) on the general theory include Weierstrass, Schwarz, Dedekind, Hölder, Berloty, Poincaré, Study, and Macfarlane.

ART. 5. QUATERNIONS AND AUSDEHNUNGSLEHRE.

Quaternions and Ausdehnungslehre‡ are so closely related to complex quantity, and the latter to complex number, that the brief sketch of their development is introduced at this point. Caspar Wessel's contributions to the theory of complex quantity and quaternions remained unnoticed in the proceedings of the Copenhagen Academy. Argand's attempts to extend his method of complex numbers beyond the space of two dimensions failed. Servois (1813), however, almost trespassed on the quaternion field. Nevertheless there were fewer traces of the theory anterior to the labors of Hamilton than is usual in the case of great discoveries. Hamilton discovered the principle of quaternions in 1843, and the next year his first contribution to the theory appeared, thus extending the Argand idea to three-dimensional space. This step neces-

* Chapman, C. H., Weierstrass and Dedekind on General Complex Numbers, in Bulletin New York Mathematical Society, Vol. I, p. 150; Study, E., Aeltere und neuere Untersuchungen über Systeme complexer Zahlen, Mathematical Papers Chicago Congress, p. 367; bibliography, p. 381.

† Klein, F., Evanston Lectures, Lect. VIII.

‡ Tait, P. G., on Quaternions, Encyclopædia Britannica; Schlegel, V., Die Grassmann'sche Ausdehnungslehre, Schlömilch's Zeitschrift, Vol. XLI.

sitated an expansion of the idea of $r(\cos\phi + j\sin\phi)$ such that while r should be a real number and ϕ a real angle, i, j, or k should be any directed unit line such that $i^2 = j^2 = k^2 = -1$. It also necessitated a withdrawal of the commutative law of multiplication, the adherence to which obstructed earlier discovery. It was not until 1853 that Hamilton's Lectures on Quarternions appeared, followed (1866) by his Elements of Quaternions.

In the same year in which Hamilton published his discovery (1844), Grassmann gave to the world his famous work, Die lineale Ausdehnungslehre, although he seems to have been in possession of the theory as early as 1840. Differing from Hamilton's Quaternions in many features, there are several essential principles held in common which each writer discovered independently of the other.*

Following Hamilton, there have appeared in Great Britain numerous papers and works by Tait (1867), Kelland and Tait (1873), Sylvester, and McAulay (1893). On the Continent Hankel (1867), Hoüel (1874), and Laisant (1877, 1881) have written on the theory, but it has attracted relatively little attention. In America, Benjamin Peirce (1870) has been especially prominent in developing the quaternion theory, and Hardy (1881) Macfarlane, and Hathaway (1896) have contributed to the subject. The difficulties have been largely in the notation. In attempting to improve this symbolism Macfarlane has aimed at showing how a space analysis can be developed embracing algebra, trigonometry, complex numbers, Grassmann's method, and quaternions, and has considered the general principles of vector and versor analysis, the versor being circular, elliptic logarithmic, or hyperbolic. Other recent contributors to the algebra of vectors are Gibbs (from 1881) and Heaviside (from 1885).

The followers of Grassmann † have not been much more

* These are set forth in a paper by J. W. Gibbs, Nature, Vol. XLIV, p. 79.

† For bibliography see Schlegel, V., Die Grassmann'sche Ausdehnungslehre, Schlömilch's Zeitschrift, Vol. XLI.

numerous than those of Hamilton. Schlegel has been one of the chief contributors in Germany, and Peano in Italy. In America, Hyde (Directional Calculus, 1890) has made a plea for the Grassmann theory.*

Along lines analogous to those of Hamilton and Grassmann have been the contributions of Scheffler. While the two former sacrificed the commutative law, Scheffler (1846, 1851, 1880) sacrificed the distributive. This sacrifice of fundamental laws has led to an investigation of the field in which these laws are valid, an investigation to which Grassmann (1872), Cayley, Ellis, Boole, Schröder (1890–91), and Kraft (1893) have contributed. Another great contribution of Cayley's along similar lines is the theory of matrices (1858).

ART. 6. THEORY OF EQUATIONS.

The Theory of Numerical Equations † concerns itself first with the location of the roots, and then with their approximation. Neither problem is new, but the first noteworthy contribution to the former in the nineteenth century was Budan's (1807). Fourier's work was undertaken at about the same time, but appeared posthumously in 1831. All processes were, however, exceedingly cumbersome until Sturm (1829) communicated to the French Academy the famous theorem which bears his name and which constitutes one of the most brilliant discoveries of algebraic analysis.

The Approximation of the Roots, once they are located, can be made by several processes. Newton (1711), for example, gave a method which Fourier perfected; and Lagrange (1767) discovered an ingenious way of expressing the root as a continued fraction, a process which Vincent (1836) elaborated. It

* For Macfarlane's Digest of views of English and American writers, see Proceedings American Association for Advancement of Science, 1891.

† Cayley, A., Equations, and Kelland, P., Algebra, in Encyclopædia Britannica; Favaro, A., Notizie storico-critiche sulla costruzione delle equazioni. Modena, 1878; Cantor, M., Geschichte der Mathematik, Vol. III, p. 375.

was, however, reserved for Horner (1819) to suggest the most practical method yet known, the one now commonly used. With Horner and Sturm this branch practically closes. The calculation of the imaginary roots by approximation is still an open field.

The Fundamental Theorem* that every numerical equation has a root was generally assumed until the latter part of the eighteenth century. D'Alembert (1746) gave a demonstration, as did Lagrange (1772), Laplace (1795), Gauss (1799) and Argand (1806). The general theorem that every algebraic equation of the nth degree has exactly n roots and no more follows as a special case of Cauchy's proposition (1831) as to the number of roots within a given contour. Proofs are also due to Gauss, Serret, Clifford (1876), Malet (1878), and many others.

The Impossibility of Expressing the Roots of an equation as algebraic functions of the coefficients when the degree exceeds 4 was anticipated by Gauss and announced by Ruffini, and the belief in the fact became strengthened by the failure of Lagrange's methods for these cases. But the first strict proof is due to Abel, whose early death cut short his labors in this and other fields.

The Quintic Equation has naturally been an object of special study. Lagrange showed that its solution depends on that of a sextic, "Lagrange's resolvent sextic," and Malfatti and Vandermonde investigated the construction of resolvents. The resolvent sextic was somewhat simplified by Cockle and Harley (1858–59) and by Cayley (1861), but Kronecker (1858) was the first to establish a resolvent by which a real simplification was effected. The transformation of the general quintic into the trinomial form $x^5 + ax + b = 0$ by the extraction of square and cube roots only, was first shown to be possible by

* Loria, Gino, Esame di alcune ricerche concernenti l'esistenza di radici nelle equazioni algebriche; Bibliotheca Mathematica, 1891, p. 99; bibliography on p. 107. Pierpont, J., On the Ruffini-Abelian theorem, Bulletin of American Mathematical Society, Vol. II, p. 200.

Bring (1786) and independently by Jerrard * (1834). Hermite (1858) actually effected this reduction, by means of Tschirnhausen's theorem, in connection with his solution by elliptic functions.

The Modern Theory of Equations may be said to date from Abel and Galois. The latter's special memoir on the subject, not published until 1846, fifteen years after his death, placed the theory on a definite base. To him is due the discovery that to each equation corresponds a group of substitutions (the "group of the equation") in which are reflected its essential characteristics.† Galois's untimely death left without sufficient demonstration several important propositions, a gap which Betti (1852) has filled. Jordan, Hermite, and Kronecker were also among the earlier ones to add to the theory. Just prior to Galois's researches Abel (1824), proceeding from the fact that a rational function of five letters having less than five values cannot have more than two, showed that the roots of a general quintic equation cannot be expressed in terms of its coefficients by means of radicals. He then investigated special forms of quintic equations which admit of solution by the extraction of a finite number of roots. Hermite, Sylvester, and Brioschi have applied the invariant theory of binary forms to the same subject.

From the point of view of the group the solution by radicals, formerly the goal of the algebraist, now appears as a single link in a long chain of questions relative to the transformation of irrationals and to their classification. Klein (1884) has handled the whole subject of the quintic equation in a simple manner by introducing the icosahedron equation as the normal form, and has shown that the method can be generalized so as to embrace the whole theory of higher equations.‡ He and Gordan (from 1879) have attacked those equations of

* Harley, R., A contribution of the history . . . of the general equation of the fifth degree, Quarterly Journal of Mathematics, Vol. VI, p. 38.

† See Art. 7.

‡ Klein, F., Vorlesungen über das Ikosaeder, 1884.

the sixth and seventh degrees which have a Galois group of 168 substitutions, Gordan performing the reduction of the equation of the seventh degree to the ternary problem. Klein (1888) has shown that the equation of the twenty-seventh degree occurring in the theory of cubic surfaces can be reduced to a normal problem in four variables, and Burkhardt (1893) has performed the reduction, the quaternary groups involved having been discussed by Maschke (from 1887).

Thus the attempt to solve the quintic equation by means of radicals has given place to their treatment by transcendents. Hermite (1858) has shown the possibility of the solution, by the use of elliptic functions, of any Bring quintic, and hence of any equation of the fifth degree. Kronecker (1858), working from a different standpoint, has reached the same results, and his method has since been simplified by Brioschi. More recently Kronecker, Gordan, Kiepert, and Klein, have contributed to the same subject, and the sextic equation has been attacked by Maschke and Brioschi through the medium of hyperelliptic functions.

Binomial Equations, reducible to the form $x^n - 1 = 0$, admit of ready solution by the familiar trigonometric formula $x = \cos\frac{2k\pi}{n} + i\sin\frac{2k\pi}{n}$; but it was reserved for Gauss (1801) to show that an algebraic solution is possible. Lagrange (1808) extended the theory, and its application to geometry is one of the leading additions of the century. Abel, generalizing Gauss's results, contributed the important theorem that if two roots of an irreducible equation are so connected that the one can be expressed rationally in terms of the other, the equation yields to radicals if the degree is prime and otherwise depends on the solution of lower equations. The binomial equation, or rather the equation $\sum_0^{n-1} x^m = 0$, is one of this class considered by Abel, and hence called (by Kronecker) Abelian Equations. The binomial equation has been treated notably by Richelot (1832), Jacobi (1837), Eisenstein (1844, 1850), Cayley (1851), and Kronecker (1854), and is the subject of a treatise by Bachmann (1872). Among the most recent writers on Abelian equations is Pellet (1891).

Certain special equations of importance in geometry have been the subject of study by Hesse, Steiner, Cayley, Clebsch, Salmon, and Kummer. Such are equations of the ninth degree determining the points of inflection of a curve of the third degree, and of the twenty-seventh degree determining the points in which a curve of the third degree can have contact of the fifth order with a conic.

Symmetric Functions of the coefficients, and those which remain unchanged through some or all of the permutations of the roots, are subjects of great importance in the present theory. The first formulas for the computation of the symmetric functions of the roots of an equation seem to have been worked out by Newton, although Girard (1629) had given, without proof, a formula for the power sum. In the eighteenth century Lagrange (1768) and Waring (1770, 1782) contributed to the theory, but the first tables, reaching to the tenth degree, appeared in 1809 in the Meyer-Hirsch Aufgabensammlung. In Cauchy's celebrated memoir on determinants (1812) the subject began to assume new prominence, and both he and Gauss (1816) made numerous and valuable contributions to the theory. It is, however, since the discoveries by Galois that the subject has become one of great importance. Cayley (1857) has given simple rules for the degree and weight of symmetric functions, and he and Brioschi have simplified the computation of tables.

Methods of Elimination and of finding the resultant (Bezout) or eliminant (De Morgan) occupied a number of eighteenth-century algebraists, prominent among them being Euler (1748), whose method, based on symmetric functions, was improved by Cramer (1750) and Bezout (1764). The leading steps in the development are represented by Lagrange (1770–71), Jacobi, Sylvester (1840), Cayley (1848, 1857), Hesse (1843, 1859), Bruno (1859), and Katter (1876). Sylvester's dialytic method appeared in 1841, and to him is also due (1851) the name and a portion of the theory of the discriminant. Among recent writers on the general theory may be mentioned Burnside and Pellet (from 1887).

ART. 7. SUBSTITUTIONS AND GROUPS.

The Theories of Substitutions and Groups* are among the most important in the whole mathematical field, the study of groups and the search for invariants now occupying the attention of many mathematicians. The first recognition of the importance of the combinatory analysis occurs in the problem of forming an mth-degree equation having for roots m of the roots of a given nth-degree equation ($m < n$). For simple cases the problem goes back to Hudde (1659). Saunderson (1740) noted that the determination of the quadratic factors of a biquadratic expression necessarily leads to a sextic equation, and Le Sœur (1748) and Waring (1762 to 1782) still further elaborated the idea.

Lagrange† first undertook a scientific treatment of the theory of substitutions. Prior to his time the various methods of solving lower equations had existed rather as isolated artifices than as a unified theory.‡ Through the great power of analysis possessed by Lagrange (1770, 1771) a common foundation was discovered, and on this was built the theory of substitutions. He undertook to examine the methods then known, and to show a priori why these succeeded below the quintic, but otherwise failed. In his investigation he discovered the important fact that the roots of all resolvents (résolvantes, réduites) which he examined are rational functions of the roots of the respective equations. To study the properties of these functions he invented a "Calcul des Combinaisons," the first

* Netto, E., Theory of Substitutions, translated by Cole; Cayley, A., Equations, Encyclopædia Britannica, 9th edition.

† Pierpont, James, Lagrange's Place in the Theory of Substitutions, Bulletin of American Mathematical Society, Vol. I, p. 196.

‡ Matthiessen, L., Grundzüge der antiken und modernen Algebra der litteralen Gleichungen, Leipzig, 1878.

important step towards a theory of substitutions. Mention should also be made of the contemporary labors of Vandermonde (1770) as foreshadowing the coming theory.

The next great step was taken by Ruffini* (1799). Beginning like Lagrange with a discussion of the methods of solving lower equations, he attempted the proof of the impossibility of solving the quintic and higher equations. While the attempt failed, it is noteworthy in that it opens with the classification of the various "permutations" of the coefficients, using the word to mean what Cauchy calls a "système des substitutions conjuguées," or simply a "système conjugué," and Galois calls a "group of substitutions." Ruffini distinguishes what are now called intransitive, transitive and imprimitive, and transitive and primitive groups, and (1801) freely uses the group of an equation under the name "l'assieme della permutazioni." He also publishes a letter from Abbati to himself, in which the group idea is prominent.

To Galois, however, the honor of establishing the theory of groups is generally awarded. He found that if $r_1, r_2, \ldots r_n$ are the n roots of an equation, there is always a group of permutations of the r's such that (1) every function of the roots invariable by the substitutions of the group is rationally known, and (2), reciprocally, every rationally determinable function of the roots is invariable by the substitutions of the group. Galois also contributed to the theory of modular equations and to that of elliptic functions. His first publication on the group theory was made at the age of eighteen (1829), but his contributions attracted little attention until the publication of his collected papers in 1846 (Liouville, Vol. XI).

Cayley and Cauchy were among the first to appreciate the importance of the theory, and to the latter especially are due a number of important theorems. The popularizing of the subject is largely due to Serret, who has devoted section IV of his

* Burkhardt, H., Die Anfänge der Gruppentheorie und Paolo Ruffini, Abhandlungen zur Geschichte der Mathematik, VI, 1892, p. 119. Italian by E. Pascal, Brioschi's Annali di Matematica, 1894.

algebra to the theory; to Camille Jordan, whose Traité des Substitutions is a classic; and to Netto (1882), whose work has been translated into English by Cole (1892). Bertrand, Hermite, Frobenius, Kronecker, and Mathieu have added to the theory. The general problem to determine the number of groups of *n* given letters still awaits solution.

But overshadowing all others in recent years in carrying on the labors of Galois and his followers in the study of discontinuous groups stand Klein, Lie, Poincaré, and Picard. Besides these discontinuous groups there are other classes, one of which, that of finite continuous groups, is especially important in the theory of differential equations. It is this class which Lie (from 1884) has studied, creating the most important of the recent departments of mathematics, the theory of transformation groups. Of value, too, have been the labors of Killing on the structure of groups, Study's application of the group theory to complex numbers, and the work of Schur and Maurer.

ART. 8. DETERMINANTS.

The Theory of Determinants* may be said to take its origin with Leibniz (1693), following whom Cramer (1750) added slightly to the theory, treating the subject, as did his predecessor, wholly in relation to sets of equations. The recurrent law was first announced by Bezout (1764). But it was Vandermonde (1771) who first recognized determinants as independent functions. To him is due the first connected exposition of the theory, and he may be called its formal founder. Laplace (1772) gave the general method of expanding a determinant in terms of its complementary minors, although Vandermonde had already given a special case. Immediately following, Lagrange (1773) treated determinants of the second

* Muir, T., Theory of Determinants in the Historical Order of its Development, Part I, 1890; Baltzer, R., Theorie und Anwendung der Determinanten. 1881. The writer is under obligations to Professor Weld, who contributes Chap. II, for valuable assistance in compiling this article.

and third order, possibly stopping here because the idea of hyperspace was not then in vogue. Although contributing nothing to the general theory, Lagrange was the first to apply determinants to questions foreign to eliminations, and to him are due many special identities which have since been brought under well-known theorems. During the next quarter of a century little of importance was done. Hindenburg (1784) and Rothe (1800) kept the subject open, but Gauss (1801) made the next advance. Like Lagrange, he made much use of determinants in the theory of numbers. He introduced the word "determinants" (Laplace had used "resultant"), though not in the present signification,* but rather as applied to the discriminant of a quantic. Gauss also arrived at the notion of reciprocal determinants, and came very near the multiplication theorem. The next contributor of importance is Binet (1811, 1812), who formally stated the theorem relating to the product of two matrices of m columns and n rows, which for the special case of $m = n$ reduces to the multiplication theorem. On the same day (Nov. 30, 1812) that Binet presented his paper to the Academy, Cauchy also presented one on the subject. In this he used the word "determinant" in its present sense, summarized and simplified what was then known on the subject, improved the notation, and gave the multiplication theorem with a proof more satisfactory than Binet's. He was the first to grasp the subject as a whole; before him there were determinants, with him begins their theory in its generality.

The next great contributor, and the greatest save Cauchy, was Jacobi (from 1827). With him the word "determinant" received its final acceptance. He early used the functional determinant which Sylvester has called the "Jacobian," and in his famous memoirs in Crelle for 1841 he specially treats this subject, as well as that class of alternating functions which Sylvester has called "Alternants." But about the time of Jacobi's closing memoirs, Sylvester (1839) and Cayley began

* "Numerum *bb–ac*, cuius indole proprietates formæ (a, b, c) imprimis pendere in sequentibus docebimus, determinantem huius uocabimus."

their great work, a work which it is impossible to briefly summarize, but which represents the development of the theory to the present time.

The study of special forms of determinants has been the natural result of the completion of the general theory. Axi-symmetric determinants have been studied by Lebesgue, Hesse, and Sylvester; per-symmetric determinants by Sylvester and Hankel; circulants by Catalan, Spottiswoode, Glaisher, and Scott; skew determinants and Pfaffians, in connection with the theory of orthogonal transformation, by Cayley; continuants by Sylvester; Wronskians (so called by Muir) by Christoffel and Frobenius; compound determinants by Sylvester, Reiss, and Picquet; Jacobians and Hessians by Sylvester; and symmetric gauche determinants by Trudi. Of the text-books on the subject Spottiswoode's was the first. In America, Hanus (1886) and Weld (1893) have published treatises.

Art. 9. Quantics.

The Theory of Quantics or Forms * appeared in embryo in the Berlin memoirs of Lagrange (1773, 1775), who considered binary quadratic forms of the type $ax^2 + bxy + cy^2$, and established the invariance of the discriminant of that type when $x + \lambda y$ is put for x. He classified forms of that type according to the sign of $b^2 - 4ac$, and introduced the ideas of transformation and equivalence. Gauss † (1801) next took up the subject, proved the invariance of the discriminants of binary and ternary quadratic forms, and systematized the theory of binary quadratic forms, a subject elaborated by H. J. S. Smith, Eisenstein, Dirichlet, Lipschitz, Poincaré, and Cayley. Galois also entered the field, in his theory of groups (1829), and

* Meyer, W. F., Bericht über den gegenwärtigen Stand der Invariantentheorie. Jahresbericht der deutschen Mathematiker-Vereinigung, Vol. I, 1890–91; Berlin 1892, p. 97. See also the review by Franklin in Bulletin New York Mathematical Society, Vol. III, p. 187; Biography of Cayley, Collected Papers, VIII, p. ix, and Proceedings of Royal Society, 1895.

† See Art. 2.

the first step towards the establishment of the distinct theory is sometimes attributed to Hesse in his investigations of the plane curve of the third order.

It is, however, to Boole (1841) that the real foundation of the theory of invariants is generally ascribed. He first showed the generality of the invariant property of the discriminant, which Lagrange and Gauss had found for special forms. Inspired by Boole's discovery Cayley took up the study in a memoir "On the Theory of Linear Transformations" (1845), which was followed (1846) by investigations concerning covariants and by the discovery of the symbolic method of finding invariants. By reason of these discoveries concerning invariants and covariants (which at first he called "hyperdeterminants") he is regarded as the founder of what is variously called Modern Algebra, Theory of Forms, Theory of Quantics, and the Theory of Invariants and Covariants. His ten memoirs on the subject began in 1854, and rank among the greatest which have ever been produced upon a single theory. Sylvester soon joined Cayley in this work, and his originality and vigor in discovery soon made both himself and the subject prominent. To him are due (1851–54) the foundations of the general theory, upon which later writers have largely built, as well as most of the terminology of the subject.

Meanwhile in Germany Eisenstein (1843) had become aware of the simplest invariants and covariants of a cubic and biquadratic form, and Hesse and Grassmann had both (1844) touched upon the subject. But it was Aronhold (1849) who first made the new theory known. He devised the symbolic method now common in Germany, discovered the invariants of a ternary cubic and their relations to the discriminant, and, with Cayley and Sylvester, studied those differential equations which are satisfied by invariants and covariants of binary quantics. His symbolic method has been carried on by Clebsch, Gordan, and more recently by Study (1889) and Stroh (1890), in lines quite different from those of the English school.

In France Hermite early took up the work (1851). He

discovered (1854) the law of reciprocity that to every covariant or invariant of degree ρ and order r of a form of the mth order corresponds also a covariant or invariant of degree m and of order r of a form of the ρth order. At the same time (1854) Brioschi joined the movement, and his contributions have been among the most valuable. Salmon's Higher Plane Curves (1852) and Higher Algebra (1859) should also be mentioned as marking an epoch in the theory.

Gordan entered the field, as a critic of Cayley, in 1868. He added greatly to the theory, especially by his theorem on the Endlichkeit des Formensystems, the proof for which has since been simplified. This theory of the finiteness of the number of invariants and covariants of a binary form has since been extended by Peano (1882), Hilbert (1884), and Mertens (1886). Hilbert (1890) succeeded in showing the finiteness of the complete systems for forms in n variables, a proof which Story has simplified.

Clebsch* did more than any other to introduce into Germany the work of Cayley and Sylvester, interpreting the projective geometry by their theory of invariants, and correlating it with Riemann's theory of functions. Especially since the publication of his work on forms (1871) the subject has attracted such scholars as Weierstrass, Kronecker, Mansion, Noether, Hilbert, Klein, Lie, Beltrami, Burkhardt, and many others. On binary forms Faà di Bruno's work is well known, as is Study's (1889) on ternary forms. De Toledo (1889) and Elliott (1895) have published treatises on the subject.

Dublin University has also furnished a considerable corps of contributors, among whom MacCullagh, Hamilton, Salmon, Michael and Ralph Roberts, and Burnside may be especially mentioned. Burnside, who wrote the latter part of Burnside and Panton's Theory of Equations, has set forth a method of transformation which is fertile in geometric interpretation and binds together binary and certain ternary forms.

* Klein's Evanston Lectures, Lect. I.

The equivalence problem of quadratic and bilinear forms has attracted the attention of Weierstrass, Kronecker, Christoffel, Frobenius, Lie, and more recently of Rosenow (Crelle, 108), Werner (1889), Killing (1890), and Scheffers (1891). The equivalence problem of non-quadratic forms has been studied by Christoffel. Schwarz (1872), Fuchs (1875–76), Klein (1877, 1884), Brioschi (1877), and Maschke (1887) have contributed to the theory of forms with linear transformations into themselves. Cayley (especially from 1870) and Sylvester (1877) have worked out the methods of denumeration by means of generating functions. Differential invariants have been studied by Sylvester, MacMahon, and Hammond. Starting from the differential invariant, which Cayley has termed the Schwarzian derivative, Sylvester (1885) has founded the theory of reciprocants, to which MacMahon, Hammond, Leudesdorf, Elliott, Forsyth, and Halphen have contributed. Canonical forms have been studied by Sylvester (1851), Cayley, and Hermite (to whom the term "canonical form" is due), and more recently by Rosanes (1873), Brill (1882), Gundelfinger (1883), and Hilbert (1886).

The Geometric Theory of Binary Forms may be traced to Poncelet and his followers. But the modern treatment has its origin in connection with the theory of elliptic modular functions, and dates from Dedekind's letter to Borchardt (Crelle, 1877). The names of Klein and Hurwitz are prominent in this connection. On the method of nets (réseaux), another geometric treatment of binary quadratic forms Gauss (1831), Dirichlet (1850), and Poincaré (1880) have written.

ART. 10. CALCULUS.

The Differential and Integral Calculus,* dating from Newton and Leibniz, was quite complete in its general range at

* Williamson, B., Infinitesimal Calculus, Encyclopædia Britannica, 9th edition; Cantor, M., Geschichte der Mathematik, Vol. III, pp. 150–316; Vivanti, G., Note sur l'histoire de l'infiniment petit, Bibliotheca Mathematica, 1894, p. 1; Mansion, P., Esquisse de l'histoire du calcul infinitésimal, Ghent, 1887. Le

the close of the eighteenth century. Aside from the study of first principles, to which Gauss, Cauchy, Jordan, Picard, Méray, and those whose names are mentioned in connection with the theory of functions, have contributed, there must be mentioned the development of symbolic methods, the theory of definite integrals, the calculus of variations, the theory of differential equations, and the numerous applications of the Newtonian calculus to physical problems. Among those who have prepared noteworthy general treatises are Cauchy (1821), Raabe (1839–47), Duhamel (1856), Sturm (1857–59), Bertrand (1864), Serret (1868), Jordan (2d ed., 1893), and Picard (1891–93). A recent contribution to analysis which promises to be valuable is Oltramare's Calcul de Généralization (1893).

Abel seems to have been the first to consider in a general way the question as to what differential expressions can be integrated in a finite form by the aid of ordinary functions, an investigation extended by Liouville. Cauchy early undertook the general theory of determining definite integrals, and the subject has been prominent during the century. Frullani's theorem (1821), Bierens de Haan's work on the theory (1862) and his elaborate tables (1867), Dirichlet's lectures (1858) embodied in Meyer's treatise (1871), and numerous memoirs of Legendre, Poisson, Plana, Raabe, Sohncke, Schlömilch, Elliott, Leudesdorf, and Kronecker are among the noteworthy contributions.

Eulerian Integrals were first studied by Euler and afterwards investigated by Legendre, by whom they were classed as Eulerian integrals of the first and second species, as follows: $\int_0^1 x^{n-1}(1-x)^{n-1}dx$, $\int_0^\infty e^{-x}x^{n-1}dx$, although these were not the exact forms of Euler's study. If n is integral, it follows that $\int_0^\infty e^{-x}x^{n-1}dx = n!$, but if n is fractional it is a transcendent function. To it Legendre assigned the symbol Γ, and it is

deux centième anniversaire de l'invention du calcul différentiel; Mathesis, Vol. IV, p. 163.

now called the gamma function. To the subject Dirichlet has contributed an important theorem (Liouville, 1839), which has been elaborated by Liouville, Catalan, Leslie Ellis, and others. On the evaluation of Γx and $\log \Gamma x$ Raabe (1843–44), Bauer (1859), and Gudermann (1845) have written. Legendre's great table appeared in 1816.

Symbolic Methods may be traced back to Taylor, and the analogy between successive differentiation and ordinary exponentials had been observed by numerous writers before the nineteenth century. Arbogast (1800) was the first, however, to separate the symbol of operation from that of quantity in a differential equation. François (1812) and Servois (1814) seem to have been the first to give correct rules on the subject. Hargreave (1848) applied these methods in his memoir on differential equations, and Boole freely employed them. Grassmann and Hankel made great use of the theory, the former in studying equations, the latter in his theory of complex numbers.

The Calculus of Variations* may be said to begin with a problem of Johann Bernoulli's (1696). It immediately occupied the attention of Jakob Bernoulli and the Marquis de l'Hôpital, but Euler first elaborated the subject. His contributions began in 1733, and his Elementa Calculi Variationum gave to the science its name. Lagrange contributed extensively to the theory, and Legendre (1786) laid down a method, not entirely satisfactory, for the discrimination of maxima and minima. To this discrimination Brunacci (1810), Gauss (1829), Poisson (1831), Ostrogradsky (1834), and Jacobi (1837) have been among the contributors. An important general work is that of Sarrus (1842) which was condensed and improved by Cauchy (1844). Other valuable treatises and memoirs have been written by Strauch (1849), Jellett (1850), Hesse (1857), Clebsch (1858), and Carll (1885), but perhaps the most

* Carll, L. B., Calculus of Variations, New York, 1885, Chap. V; Todhunter, I., History of the Progress of the Calculus of Variations, London, 1861; Reiff, R., Die Anfänge der Variationsrechnung, Mathematisch-naturwissenschaftliche Mittheilungen, Tübingen, 1887, p. 90.

important work of the century is that of Weierstrass. His celebrated course on the theory is epoch-making, and it may be asserted that he was the first to place it on a firm and unquestionable foundation.

The Application of the Infinitesimal Calculus to problems in physics and astronomy was contemporary with the origin of the science. All through the eighteenth century these applications were multiplied, until at its close Laplace and Lagrange had brought the whole range of the study of forces into the realm of analysis. To Lagrange (1773) we owe the introduction of the theory of the potential* into dynamics, although the name "potential function" and the fundamental memoir of the subject are due to Green (1827, printed in 1828). The name "potential" is due to Gauss (1840), and the distinction between potential and potential function to Clausius. With its development are connected the names of Dirichlet, Riemann, Neumann, Heine, Kronecker, Lipschitz, Christoffel, Kirchhoff, Beltrami, and many of the leading physicists of the century.

It is impossible in this place to enter into the great variety of other applications of analysis to physical problems. Among them are the investigations of Euler on vibrating chords; Sophie Germain on elastic membranes; Poisson, Lamé, Saint-Venant, and Clebsch on the elasticity of three-dimensional bodies; Fourier on heat diffusion; Fresnel on light; Maxwell, Helmholtz, and Hertz on electricity; Hansen, Hill, and Gyldén on astronomy; Maxwell on spherical harmonics; Lord Rayleigh on acoustics; and the contributions of Dirichlet, Weber, Kirchhoff, F. Neumann, Lord Kelvin, Clausius, Bjerknes, MacCullagh, and Fuhrmann to physics in general. The labors of Helmholtz should be especially mentioned, since he contributed to the theories of dynamics, electricity, etc., and brought his great analytical powers to bear on the fundamental axioms of mechanics as well as on those of pure mathematics.

* Bacharach, M., Abriss der Geschichte der Potentialtheorie, 1883. This contains an extensive bibliography.

ART. 11. DIFFERENTIAL EQUATIONS.

The Theory of Differential Equations * has been called by Lie † the most important of modern mathematics. The influence of geometry, physics, and astronomy, starting with Newton and Leibniz, and further manifested through the Bernoullis, Riccati, and Clairaut, but chiefly through d'Alembert and Euler, has been very marked, and especially on the theory of linear partial differential equations with constant coefficients. The first method of integrating linear ordinary differential equations with constant coefficients is due to Euler, who made the solution of his type, $\frac{d^ny}{dx^n} + A_1\frac{d^{n-1}y}{dx^{n-1}} + \ldots + A_n y = 0$, depend on that of the algebraic equation of the nth degree, $F(z) = z^n + A_1 z^{n-1} + \ldots + A_n = 0$, in which z^k takes the place of $\frac{d^ky}{dx^k}$ $(k = 1, 2, \ldots n)$. This equation $F(z) = 0$, is the "characteristic" equation considered later by Monge and Cauchy.

The theory of linear partial differential equations may be said to begin with Lagrange (1779 to 1785). Monge (1809) treated ordinary and partial differential equations of the first and second order, uniting the theory to geometry, and introducing the notion of the "characteristic," the curve represented by $F(z) = 0$, which has recently been investigated by Darboux,

* Cantor, M., Geschichte der Mathematik, Vol. III, p. 429; Schlesinger, L., Handbuch der Theorie der linearen Differentialgleichungen, Vol. I, 1895, an excellent historical view; review by Mathews in Nature, Vol. LII, p. 313; Lie, S., Zur allgemeinen Theorie der partiellen Differentialgleichungen, Berichte über die Verhandlungen der Gesellschaft der Wissenschaften zu Leipzig, 1895; Mansion, P., Theorie der partiellen Differentialgleichungen 1er Ordnung, German by Maser, Leipzig, 1892, excellent on history; Craig, T., Some of the Developments in the Theory of Ordinary Differential Equations, 1878–1893, Bulletin New York Mathematical Society, Vol. II, p. 119; Goursat, E., Leçons sur l'intégration des équations aux dérivées partielles du premier ordre, Paris, 1891; Burkhardt, H., and Heffter, L., in Mathematical Papers of Chicago Congress, p. 13 and p. 96.

† "In der ganzen modernen Mathematik ist die Theorie der Differentialgleichungen die wichtigste Disciplin"

Levy, and Lie. Pfaff (1814, 1815) gave the first general method of integrating partial differential equations of the first order, a method of which Gauss (1815) at once recognized the value and of which he gave an analysis. Soon after, Cauchy (1819) gave a simpler method, attacking the subject from the analytical standpoint, but using the Monge characteristic. To him is also due the theorem, corresponding to the fundamental theorem of algebra, that every differential equation defines a function expressible by means of a convergent series, a proposition more simply proved by Briot and Bouquet, and also by Picard (1891). Jacobi (1827) also gave an analysis of Pfaff's method, besides developing an original one (1836) which Clebsch published (1862). Clebsch's own method appeared in 1866, and others are due to Boole (1859), Korkine (1869), and A. Mayer (1872). Pfaff's problem has been a prominent subject of investigation, and with it are connected the names of Natani (1859), Clebsch (1861, 1862), DuBois-Reymond (1869), Cayley, Baltzer, Frobenius, Morera, Darboux, and Lie. The next great improvement in the theory of partial differential equations of the first order is due to Lie (1872), by whom the whole subject has been placed on a rigid foundation. Since about 1870, Darboux, Kovalevsky, Méray, Mansion, Graindorge, and Imschenetsky have been prominent in this line. The theory of partial differential equations of the second and higher orders, beginning with Laplace and Monge, was notably advanced by Ampère (1840). Imschenetsky * has summarized the contributions to 1873, but the theory remains in an imperfect state.

The integration of partial differential equations with three or more variables was the object of elaborate investigations by Lagrange, and his name is still connected with certain subsidiary equations. To him and to Charpit, who did much to develop the theory, is due one of the methods for integrating the general equation with two variables, a method which now bears Charpit's name.

* Grunert's Archiv für Mathematik, Vol. LIV.

The theory of singular solutions of ordinary and partial differential equations has been a subject of research from the time of Leibniz, but only since the middle of the present century has it received especial attention. A valuable but little-known work on the subject is that of Houtain (1854). Darboux (from 1873) has been a leader in the theory, and in the geometric interpretation of these solutions he has opened a field which has been worked by various writers, notably Casorati and Cayley. To the latter is due (1872) the theory of singular solutions of differential equations of the first order as at present accepted.

The primitive attempt in dealing with differential equations had in view a reduction to quadratures. As it had been the hope of eighteenth-century algebraists to find a method for solving the general equation of the nth degree, so it was the hope of analysts to find a general method for integrating any differential equation. Gauss (1799) showed, however, that the differential equation meets its limitations very soon unless complex numbers are introduced. Hence analysts began to substitute the study of functions, thus opening a new and fertile field. Cauchy was the first to appreciate the importance of this view, and the modern theory may be said to begin with him. Thereafter the real question was to be, not whether a solution is possible by means of known functions or their integrals, but whether a given differential equation suffices for the definition of a function of the independent variable or variables, and if so, what are the characteristic properties of this function.

Within a half-century the theory of ordinary differential equations has come to be one of the most important branches of analysis, the theory of partial differential equations remaining as one still to be perfected. The difficulties of the general problem of integration are so manifest that all classes of investigators have confined themselves to the properties of the integrals in the neighborhood of certain given points. The new departure took its greatest inspiration from two memoirs by

Fuchs (Crelle, 1866, 1868), a work elaborated by Thomé and Frobenius. Collet has been a prominent contributor since 1869, although his method for integrating a non-linear system was communicated to Bertrand in 1868. Clebsch * (1873) attacked the theory along lines parallel to those followed in his theory of Abelian integrals. As the latter can be classified according to the properties of the fundamental curve which remains unchanged under a rational transformation, so Clebsch proposed to classify the transcendent functions defined by the differential equations according to the invariant properties of the corresponding surfaces $f = 0$ under rational one-to-one transformations.

Since 1870 Lie's † labors have put the entire theory of differential equations on a more satisfactory foundation. He has shown that the integration theories of the older mathematicians, which had been looked upon as isolated, can by the introduction of the concept of continuous groups of transformations be referred to a common source, and that ordinary differential equations which admit the same infinitesimal transformations present like difficulties of integration. He has also emphasized the subject of transformations of contact (Berührungstransformationen) which underlies so much of the recent theory. The modern school has also turned its attention to the theory of differential invariants, one of fundamental importance and one which Lie has made prominent. With this theory are associated the names of Cayley, Cockle, Sylvester, Forsyth, Laguerre, and Halphen. Recent writers have shown the same tendency noticeable in the work of Monge and Cauchy, the tendency to separate into two schools, the one inclining to use the geometric diagram, and represented by Schwarz, Klein, and Goursat, the other adhering to pure analysis, of which Weierstrass, Fuchs, and Frobenius are types. The work of Fuchs and the theory of elementary divisors have formed the basis of a late work by Sauvage (1895). Poincaré's

* Klein's Evanston Lectures, Lect. I.

† Klein's Evanston Lectures, Lect. II, III.

recent contributions are also very notable. His theory of Fuchsian equations (also investigated by Klein) is connected with the general theory. He has also brought the whole subject into close relations with the theory of functions. Appell has recently contributed to the theory of linear differential equations transformable into themselves by change of the function and the variable. Helge von Koch has written on infinite determinants and linear differential equations. Picard has undertaken the generalization of the work of Fuchs and Poincaré in the case of differential equations of the second order. Fabry (1885) has generalized the normal integrals of Thomé, integrals which Poincaré has called "intégrales anormales," and which Picard has recently studied. Riquier has treated the question of the existence of integrals in any differential system and given a brief summary of the history to 1895.* The number of contributors in recent times is very great, and includes, besides those already mentioned, the names of Brioschi, Königsberger, Peano, Graf, Hamburger, Graindorge, Schläfli, Glaisher, Lommel, Gilbert, Fabry, Craig, and Autonne.

ART. 12. INFINITE SERIES.

The Theory of Infinite Series† in its historical development has been divided by Reiff into three periods: (1) the period of Newton and Leibniz, that of its introduction; (2) that of Euler, the formal period; (3) the modern, that of the scientific investigation of the validity of infinite series, a period beginning with Gauss. This critical period begins with the publication of Gauss's celebrated memoir on the series $1 + \frac{\alpha . \beta}{1 . \gamma}x + \frac{\alpha . (\alpha + 1) . \beta . (\beta + 1)}{1 . 2 . \gamma . (\gamma + 1)}x^2 + \ldots$, in 1812. Euler

* Riquier, C., Mémoire sur l'existence des intégrales dans un système différentiel quelconque, etc. Mémoires des Savants étrangers, Vol. XXXII, No. 3.

† Cantor, M., Geschichte der Mathematik, Vol. III, pp. 53, 71 ; Reiff, R., Geschichte der unendlichen Reihen, Tübingen, 1889 ; Cajori, F., Bulletin New York Mathematical Society, Vol. I, p. 184; History of Teaching of Mathematics in United States, p. 361.

had already considered this series, but Gauss was the first to master it, and under the name "hypergeometric series" (due to Pfaff) it has since occupied the attention of Jacobi, Kummer, Schwarz, Cayley, Goursat, and numerous others. The particular series is not so important as is the standard of criticism which Gauss set up, embodying the simpler criteria of convergence and the questions of remainders and the range of convergence.

Gauss's contributions were not at once appreciated, and the next to call attention to the subject was Cauchy (1821), who may be considered the founder of the theory of convergence and divergence of series. He was one of the first to insist on strict tests of convergence; he showed that if two series are convergent their product is not necessarily so; and with him begins the discovery of effective criteria of convergence and divergence. It should be mentioned, however, that these terms had been introduced long before by Gregory (1668), that Euler and Gauss had given various criteria, and that Maclaurin had anticipated a few of Cauchy's discoveries. Cauchy advanced the theory of power series by his expansion of a complex function in such a form. His test for convergence is still one of the most satisfactory when the integration involved is possible.

Abel was the next important contributor. In his memoir (1826) on the series $1 + \frac{m}{1}x + \frac{m(m-1)}{2!}x^2 + \ldots$ he corrected certain of Cauchy's conclusions, and gave a completely scientific summation of the series for complex values of m and x. He was emphatic against the reckless use of series, and showed the necessity of considering the subject of continuity in questions of convergence.

Cauchy's methods led to special rather than general criteria, and the same may be said of Raabe (1832), who made the first elaborate investigation of the subject, of De Morgan (from 1842), whose logarithmic test DuBois-Reymond (1873) and Pringsheim (1889) have shown to fail within a certain region:

of Bertrand (1842), Bonnet (1843), Malmsten (1846, 1847, the latter without integration); Stokes (1847), Paucker (1852), Tchébichef (1852), and Arndt 1853). General criteria began with Kummer (1835), and have been studied by Eisenstein (1847), Weierstrass in his various contributions to the theory of functions, Dini (1867), DuBois-Reymond (1873), and many others. Pringsheim's (from 1889) memoirs present the most complete general theory.

The Theory of Uniform Convergence was treated by Cauchy (1821), his limitations being pointed out by Abel, but the first to attack it successfully were Stokes and Seidel (1847–48). Cauchy took up the problem again (1853), acknowledging Abel's criticism, and reaching the same conclusions which Stokes had already found. Thomé used the doctrine (1866), but there was great delay in recognizing the importance of distinguishing between uniform and non-uniform convergence, in spite of the demands of the theory of functions.

Semi-Convergent Series were studied by Poisson (1823), who also gave a general form for the remainder of the Maclaurin formula. The most important solution of the problem is due, however, to Jacobi (1834), who attacked the question of the remainder from a different standpoint and reached a different formula. This expression was also worked out, and another one given, by Malmsten (1847). Schlömilch (Zeitschrift, Vol. I, p. 192, 1856) also improved Jacobi's remainder, and showed the relation between the remainder and Bernoulli's function $F(x) = 1^n + 2^n + \ldots + (x-1)^n$. Genocchi (1852) has further contributed to the theory.

Among the early writers was Wronski, whose "loi suprême" (1815) was hardly recognized until Cayley (1873) brought it into prominence. Transon (1874), Ch. Lagrange (1884), Echols, and Dickstein * have published of late various memoirs on the subject.

Interpolation Formulas have been given by various writers

* Bibliotheca Mathematica, 1892–94; historical.

from Newton to the present time. Lagrange's theorem is well known, although Euler had already given an analogous form, as are also Olivier's formula (1827), and those of Minding (1830), Cauchy (1837), Jacobi (1845), Grunert (1850, 1853), Christoffel (1858), and Mehler (1864).

Fourier's Series* were being investigated as the result of physical considerations at the same time that Gauss, Abel, and Cauchy were working out the theory of infinite series. Series for the expansion of sines and cosines, of multiple arcs in powers of the sine and cosine of the arc had been treated by Jakob Bernoulli (1702) and his brother Johann (1701) and still earlier by Viète. Euler and Lagrange had simplified the subject, as have, more recently, Poinsot, Schröter, Glaisher, and Kummer. Fourier (1807) set for himself a different problem, to expand a given function of x in terms of the sines or cosines of multiples of x, a problem which he embodied in his Théorie analytique de la Chaleur (1822). Euler had already given the formulas for determining the coefficients in the series; and Lagrange had passed over them without recognizing their value, but Fourier was the first to assert and attempt to prove the general theorem. Poisson (1820–23) also attacked the problem from a different standpoint. Fourier did not, however, settle the question of convergence of his series, a matter left for Cauchy (1826) to attempt and for Dirichlet (1829) to handle in a thoroughly scientific manner. Dirichlet's treatment (Crelle, 1829), while bringing the theory of trigonometric series to a temporary conclusion, has been the subject of criticism and improvement by Riemann (1854), Heine, Lipschitz, Schläfli, and DuBois-Reymond. Among other prominent contributors to the theory of trigonometric and Fourier series have been Dini, Hermite, Halphen, Krause, Byerly and Appell.

* Historical Summary by Bôcher, Chap. IX of Byerly's Fourier's Series and Spherical Harmonics, Boston, 1893; Sachse, A., Essai historique sur la représentation d'une fonction . . . . par une série trigonométrique. Bulletin des Sciences mathématiques, Part I, 1880, pp. 43, 83.

ART. 13. THEORY OF FUNCTIONS.

The Theory of Functions * may be said to have its first development in Newton's works, although algebraists had already become familiar with irrational functions in considering cubic and quartic equations. Newton seems first to have grasped the idea of such expressions in his consideration of symmetric functions of the roots of an equation. The word was employed by Leibniz (1694), but in connection with the Cartesian geometry. In its modern sense it seems to have been first used by Johann Bernoulli, who distinguished between algebraic and transcendent functions. He also used (1718) the function symbol ϕ. Clairaut (1734) used Πx, Φx, Δx, for various functions of x, a symbolism substantially followed by d'Alembert (1747) and Euler (1753). Lagrange (1772, 1797, 1806) laid the foundations for the general theory, giving to the symbol a broader meaning, and to the symbols $f, \phi, F, \ldots$, $f', \phi', F', \ldots$ their modern signification. Gauss contributed to the theory, especially in his proofs of the fundamental theorem of algebra, and discussed and gave name to the theory of "conforme Abbildung," the "orthomorphosis" of Cayley.

Making Lagrange's work a point of departure, Cauchy so greatly developed the theory that he is justly considered one of its founders. His memoirs extend over the period 1814–1851, and cover subjects like those of integrals with imaginary limits, infinite series and questions of convergence, the application of the infinitesimal calculus to the theory of complex

* Brill, A., and Noether, M., Die Entwickelung der Theorie der algebraischen Functionen in älterer und neuerer Zeit, Bericht erstattet der Deutschen Mathematiker-Vereinigung, Jahresbericht, Vol. II, pp. 107–566, Berlin, 1894; Königsberger, L., Zur Geschichte der Theorie der elliptischen Transcendenten in den Jahren 1826–29, Leipzig, 1879; Williamson, B., Infinitesimal Calculus, Encyclopædia Britannica; Schlesinger, L., Differentialgleichungen, Vol. I, 1895; Casorati, F., Teorica delle funzioni di variabili complesse, Vol. I, 1868; Klein's Evanston Lectures. For bibliography and historical notes, see Harkness and Morley's Theory of Functions, 1893, and Forsyth's Theory of Functions, 1893; Eneström, G., Note historique sur les symboles. . . . Bibliotheca Mathematica, 1891, p 89.

numbers, the investigation of the fundamental laws of mathematics, and numerous other lines which appear in the general theory of functions as considered to-day. Originally opposed to the movement started by Gauss, the free use of complex numbers, he finally became, like Abel, its advocate. To him is largely due the present orientation of mathematical research, making prominent the theory of functions, distinguishing between classes of functions, and placing the whole subject upon a rigid foundation. The historical development of the general theory now becomes so interwoven with that of special classes of functions, and notably the elliptic and Abelian, that economy of space requires their treatment together, and hence a digression at this point.

The Theory of Elliptic Functions* is usually referred for its origin to Landen's (1775) substitution of two elliptic arcs for a single hyperbolic arc. But Jakob Bernoulli (1691) had suggested the idea of comparing non-congruent arcs of the same curve, and Johann had followed up the investigation. Fagnano (1716) had made similar studies, and both Maclaurin (1742) and d'Alembert (1746) had come upon the borderland of elliptic functions. Euler (from 1761) had summarized and extended the rudimentary theory, showing the necessity for a convenient notation for elliptic arcs, and prophesying (1766) that "such signs will afford a new sort of calculus of which I have here attempted the exposition of the first elements." Euler's investigations continued until about the time of his death (1783), and to him Legendre attributes the foundation of the theory. Euler was probably never aware of Landen's discovery.

It is to Legendre, however, that the theory of elliptic functions is largely due, and on it his fame to a considerable degree depends. His earlier treatment (1786) almost entirely substitutes a strict analytic for the geometric method. For forty years he had the theory in hand, his labor culminating in his

* Enneper, A., Elliptische Funktionen, Theorie und Geschichte, Halle, 1890; Königsberger, L., Zur Geschichte der Theorie der elliptischen Transcendenten in den Jahren 1826–29, Leipzig, 1879.

Traité des Fonctions elliptiques et des Intégrales Eulériennes (1825–28). A surprise now awaiting him is best told in his own words: "Hardly had my work seen the light—its name could scarcely have become known to scientific foreigners,—when I learned with equal surprise and satisfaction that two young mathematicians, MM. Jacobi of Königsberg and Abel of Christiania, had succeeded by their own studies in perfecting considerably the theory of elliptic functions in its highest parts." Abel began his contributions to the theory in 1825, and even then was in possession of his fundamental theorem which he communicated to the Paris Academy in 1826. This communication being so poorly transcribed was not published in full until 1841, although the theorem was sent to Crelle (1829) just before Abel's early death. Abel discovered the double periodicity of elliptic functions, and with him began the treatment of the elliptic integral as a function of the amplitude.

Jacobi, as also Legendre and Gauss, was especially cordial in praise of the delayed theorem of the youthful Abel. He calls it a "monumentum ære perennius," and his name "das Abel'sche Theorem" has since attached to it. The functions of multiple periodicity to which it refers have been called Abelian Functions. Abel's work was early proved and elucidated by Liouville and Hermite. Serret and Chasles in the Comptes Rendus, Weierstrass (1853), Clebsch and Gordan in their Theorie der Abel'schen Functionen (1866), and Briot and Bouquet in their two treatises have greatly elaborated the theory. Riemann's * (1857) celebrated memoir in Crelle presented the subject in such a novel form that his treatment was slow of acceptance. He based the theory of Abelian integrals and their inverse, the Abelian functions, on the idea of the surface now so well known by his name, and on the corresponding fundamental existence theorems. Clebsch, starting from

* Klein, Evanston Lectures, p. 3; Riemann and Modern Mathematics, translated by Ziwet, Bulletin of American Mathematical Society, Vol. I, p. 165; Burkhardt, H., Vortrag über Riemann, Göttingen, 1892.

an algebraic curve defined by its equation, made the subject more accessible, and generalized the theory of Abelian integrals to a theory of algebraic functions with several variables, thus creating a branch which has been developed by Noether, Picard, and Poincaré. The introduction of the theory of invariants and projective geometry into the domain of hyperelliptic and Abelian functions is an extension of Clebsch's scheme. In this extension, as in the general theory of Abelian functions, Klein has been a leader. With the development of the theory of Abelian functions is connected a long list of names, including those of Schottky, Humbert, C. Neumann, Fricke, Königsberger, Prym, Schwarz, Painlevé, Hurwitz, Brioschi, Borchardt, Cayley, Forsyth, and Rosenhain, besides others already mentioned.

Returning to the theory of elliptic functions, Jacobi (1827) began by adding greatly to Legendre's work. He created a new notation and gave name to the "modular equations" of which he made use. Among those who have written treatises upon the elliptic-function theory are Briot and Bouquet, Laurent, Halphen, Königsberger, Hermite, Durège, and Cayley. The introduction of the subject into the Cambridge Tripos (1873), and the fact that Cayley's only book was devoted to it, have tended to popularize the theory in England.

The Theory of Theta Functions was the simultaneous and independent creation of Jacobi and Abel (1828). Gauss's notes show that he was aware of the properties of the theta functions twenty years earlier, but he never published his investigations. Among the leading contributors to the theory are Rosenhain (1846, published in 1851) and Göpel (1847), who connected the double theta functions with the theory of Abelian functions of two variables and established the theory of hyperelliptic functions in a manner corresponding to the Jacobian theory of elliptic functions. Weierstrass has also developed the theory of theta functions independently of the form of their space boundaries, researches elaborated by Königsberger (1865) to give the addition theorem. Riemann has completed the investigation of the relation between the theory of the theta and the Abelian functions, and has raised theta functions to their present position by making them an essential part of his theory of Abelian integrals. H. J. S. Smith has included among his contributions to this subject the theory of omega functions. Among the recent contributors are Krazer and Prym (1892), and Wirtinger (1895).

Cayley was a prominent contributor to the theory of periodic functions. His memoir (1845) on doubly periodic functions extended Abel's investigations on doubly infinite products. Euler had given singly infinite products for sin x, cos x, and Abel had generalized these, obtaining for the elementary doubly periodic functions expressions for sn x, cn x, dn x. Starting from these expressions of Abel's Cayley laid a complete foundation for his theory of elliptic functions. Eisenstein (1847) followed, giving a discussion from the standpoint of pure analysis, of a general doubly infinite product, and his labors, as supplemented by Weierstrass, are classic.

The General Theory of Functions has received its present form largely from the works of Cauchy, Riemann, and Weierstrass. Endeavoring to subject all natural laws to interpretation by mathematical formulas, Riemann borrowed his methods from the theory of the potential, and found his inspiration in the contemplation of mathematics from the standpoint of the concrete. Weierstrass, on the other hand, proceeded from the purely analytic point of view. To Riemann* is due the idea of making certain partial differential equations, which express the fundamental properties of all functions, the foundation of a general analytical theory, and of seeking criteria for the determination of an analytic function by its discontinuities and boundary conditions. His theory has been elaborated by Klein (1882, and frequent memoirs) who has materially extended the theory of Riemann's surfaces. Clebsch, Lüroth, and later writers have based on this theory their researches on

* Klein, F., Riemann and Modern Mathematics, translated by Ziwet, Bulletin of American Mathematical Society, Vol. I, p. 165.

loops. Riemann's speculations were not without weak points, and these have been fortified in connection with the theory of the potential by C. Neumann, and from the analytic standpoint by Schwarz.

In both the theory of general and of elliptic and other functions, Clebsch was prominent. He introduced the systematic consideration of algebraic curves of deficiency 1, bringing to bear on the theory of elliptic functions the ideas of modern projective geometry. This theory Klein has generalized in his Theorie der elliptischen Modulfunctionen, and has extended the method to the theory of hyperelliptic and Abelian functions.

Following Riemann came the equally fundamental and original and more rigorously worked out theory of Weierstrass. His early lectures on functions are justly considered a landmark in modern mathematical development. In particular, his researches on Abelian transcendents are perhaps the most important since those of Abel and Jacobi. His contributions to the theory of elliptic functions, including the introduction of the function $\wp(u)$, are also of great importance. His contributions to the general function theory include much of the symbolism and nomenclature, and many theorems. He first announced (1866) the existence of natural limits for analytic functions, a subject further investigated by Schwarz, Klein, and Fricke. He developed the theory of functions of complex variables from its foundations, and his contributions to the theory of functions of real variables were no less marked.

Fuchs has been a prominent contributor, in particular (1872) on the general form of a function with essential singularities. On functions with an infinite number of essential singularities Mittag-Leffler (from 1882) has written and contributed a fundamental theorem. On the classification of singularities of functions Guichard (1883) has summarized and extended the researches, and Mittag-Leffler and G. Cantor have contributed to the same result. Laguerre (from 1882) was the first to discuss the "class" of transcendent functions, a subject to which Poincaré, Cesaro, Vivanti, and Hermite have also contributed. Automorphic functions, as named by Klein, have been investigated chiefly by Poincaré, who has established their general classification. The contributors to the theory include Schwarz, Fuchs, Cayley, Weber, Schlesinger, and Burnside.

The Theory of Elliptic Modular Functions, proceeding from Eisenstein's memoir (1847) and the lectures of Weierstrass on elliptic functions, has of late assumed prominence through the influence of the Klein school. Schläfli (1870), and later Klein, Dyck, Gierster, and Hurwitz, have worked out the theory which Klein and Fricke have embodied in the recent Vorlesungen über die Theorie der elliptischen Modulfunctionen (1890–92). In this theory the memoirs of Dedekind (1877), Klein (1878), and Poincaré (from 1881) have been among the most prominent.

For the names of the leading contributors to the general and special theories, including among others Jordan, Hermite, Hölder, Picard, Biermann, Darboux, Pellet, Reichardt, Burkhardt, Krause, and Humbert, reference must be had to the Brill-Noether Bericht.

Of the various special algebraïc functions space allows mention of but one class,that bearing Bessel's name. Bessel's functions * of the zero order are found in memoirs of Daniel Bernoulli (1732) and Euler (1764), and before the end of the eighteenth century all the Bessel functions of the first kind and integral order had been used. Their prominence as special functions is due, however, to Bessel (1816–17), who put them in their present form in 1824. Lagrange's series (1770), with Laplace's extension (1777), had been regarded as the best method of solving Kepler's problem (to express the variable quantities in undisturbed planetary motion in terms of the time or mean anomaly), and to improve this method Bessel's functions were first prominently used. Hankel (1869), Lommel (from 1868), F. Neumann, Heine, Graf (1893), Gray and

* Bôcher, M., A bit of mathematical history, Bulletin of New York Mathematical Society, Vol. II, p. 107.

Mathews (1895), and others have contributed to the theory. Lord Rayleigh (1878) has shown the relation between Bessel's and Laplace's functions, but they are nevertheless looked upon as a distinct system of transcendents. Tables of Bessel's functions were prepared by Bessel (1824), by Hansen (1843), and by Meissel (1888).

ART. 14. PROBABILITIES AND LEAST SQUARES.

The Theory of Probabilities and Errors* is, as applied to observations, largely a nineteenth-century development. The doctrine of probabilities dates, however, as far back as Fermat and Pascal (1654). Huygens (1657) gave the first scientific treatment of the subject, and Jakob Bernoulli's Ars Conjectandi (posthumous, 1713) and De Moivre's Doctrine of Chances (1718)† raised the subject to the plane of a branch of mathematics. The theory of errors may be traced back to Cotes's Opera Miscellanea (posthumous, 1722), but a memoir prepared by Simpson in 1755 (printed 1756) first applied the theory to the discussion of errors of observation. The reprint (1757) of this memoir lays down the axioms that positive and negative errors are equally probable, and that there are certain assignable limits within which all errors may be supposed to fall; continuous errors are discussed and a probability curve is given. Laplace (1774) made the first attempt to deduce a rule for the combination of observations from the principles of the theory of probabilities. He represented the law of probability of errors by a curve $y = \phi(x)$, x being any error and y its probability, and laid down three properties of this curve: (1) It is symmetric as to the y-axis; (2) the x-axis is an asymptote, the probability of the error ∞ being 0; (3) the area enclosed is 1, it being certain that an error exists. He deduced a formula

* Merriman, M., Method of Least Squares, New York, 1884, p. 182; Transactions of Connecticut Academy, 1877, Vol. IV, p. 151, with complete bibliography; Todhunter, I., History of the Mathematical Theory of Probability, 1865; Cantor, M., Geschichte der Mathematik, Vol. III, p. 316.

† Eneström, G., Review of Cantor, Bibliotheca Mathematica, 1896, p. 20.

for the mean of three observations. He also gave (1781) a formula for the law of facility of error (a term due to Lagrange, 1774), but one which led to unmanageable equations. Daniel Bernoulli (1778) introduced the principle of the maximum product of the probabilities of a system of concurrent errors.

The Method of Least Squares is due to Legendre (1805), who introduced it in his Nouvelles méthodes pour la détermination des orbites des comètes. In ignorance of Legendre's contribution, an Irish-American writer, Adrain, editor of "The Analyst" (1808), first deduced the law of facility of error, $\phi(x) = ce^{-h^2x^2}$, c and h being constants depending on precision of observation. He gave two proofs, the second being essentially the same as Herschel's (1850). Gauss gave the first proof which seems to have been known in Europe (the third after Adrain's) in 1809. To him is due much of the honor of placing the subject before the mathematical world, both as to the theory and its applications.

Further proofs were given by Laplace (1810, 1812), Gauss (1823), Ivory (1825, 1826), Hagen (1837), Bessel (1838), Donkin (1844, 1856), and Crofton (1870). Other contributors have been Ellis (1844), De Morgan (1864), Glaisher (1872), and Schiaparelli (1875). Peters's (1856) formula for r, the probable error of a single observation, is well known.*

Among the contributors to the general theory of probabilities in the nineteenth century have been Laplace, Lacroix (1816), Littrow (1833), Quetelet (1853), Dedekind (1860), Helmert (1872), Laurent (1873), Liagre, Didion, and Pearson. De Morgan and Boole improved the theory, but added little that was fundamentally new. Czuber has done much both in his own contributions (1884, 1891), and in his translation (1879) of Meyer. On the geometric side the influence of Miller and The Educational Times has been marked, as also that of such contributors to this journal as Crofton, McColl, Wolstenholme, Watson, and Artemas Martin.

* Bulletin of New York Mathematical Society, Vol. II, p. 57.

ART. 15. ANALYTIC GEOMETRY.

The History of Geometry* may be roughly divided into the four periods: (1) The synthetic geometry of the Greeks, practically closing with Archimedes; (2) The birth of analytic geometry, in which the synthetic geometry of Guldin, Desargues, Kepler, and Roberval merged into the coordinate geometry of Descartes and Fermat; (3) 1650 to 1800, characterized by the application of the calculus to geometry, and including the names of Newton, Leibnitz, the Bernoullis, Clairaut, Maclaurin, Euler, and Lagrange, each an analyst rather than a geometer; (4) The nineteenth century, the renaissance of pure geometry, characterized by the descriptive geometry of Monge, the modern synthetic of Poncelet, Steiner, von Staudt, and Cremona, the modern analytic founded by Plücker, the non-Euclidean hypothesis of Lobachevsky and Bolyai, and the more elementary geometry of the triangle founded by Lemoine. It is quite impossible to draw the line between the analytic and the synthetic geometry of the nineteenth century, in their historical development, and Arts. 15 and 16 should be read together.

The Analytic Geometry which Descartes gave to the world in 1637 was confined to plane curves, and the various important properties common to all algebraic curves were soon discovered. To the theory Newton contributed three celebrated theorems on the Enumeratio linearum tertii ordinis † (1706), while others are due to Cotes (1722), Maclaurin, and Waring (1762, 1772,

* Loria, G., Il passato e il presente delle principali teorie geometriche. Memorie Accademia Torino, 1887; translated into German by F. Schütte under the title Die hauptsächlichsten Theorien der Geometrie in ihrer früheren und heutigen Entwickelung, Leipzig, 1888; Chasles, M., Aperçu historique sur l'origine et le développement des méthodes en Géométrie, 1889; Chasles, M., Rapport sur les Progrès de la Géométrie, Paris, 1870; Cayley, A., Curves, Encyclopædia Britannica; Klein, F., Evanston Lectures on Mathematics, New York, 1894; A. V. Braunmühl, Historische Studie über die organische Erzeugung ebener Curven, Dyck's Katalog mathematischer Modelle, 1892.

† Ball, W. W. R., On Newton's classification of cubic curves. Transactions of London Mathematical Society, 1891, p. 104.

etc.). The scientific foundations of the theory of plane curves may be ascribed, however, to Euler (1748) and Cramer (1750). Euler distinguished between algebraic and transcendent curves, and attempted a classification of the former. Cramer is well known for the "paradox" which bears his name, an obstacle which Lamé (1818) finally removed from the theory. To Cramer is also due an attempt to put the theory of singularities of algebraic curves on a scientific foundation, although in a modern geometric sense the theory was first treated by Poncelet.

Meanwhile the study of surfaces was becoming prominent. Descartes had suggested that his geometry could be extended to three-dimensional space, Wren (1669) had discovered the two systems of generating lines on the hyperboloid of one sheet, and Parent (1700) had referred a surface to three coordinate planes. The geometry of three dimensions began to assume definite shape, however, in a memoir of Clairaut's (1731), in which, at the age of sixteen, he solved with rare elegance many of the problems relating to curves of double curvature. Euler (1760) laid the foundations for the analytic theory of curvature of surfaces, attempting the classification of those of the second degree as the ancients had classified curves of the second order. Monge, Hachette, and other members of that school entered into the study of surfaces with great zeal. Monge introduced the notion of families of surfaces, and discovered the relation between the theory of surfaces and the integration of partial differential equations, enabling each to be advantageously viewed from the standpoint of the other. The theory of surfaces has attracted a long list of contributors in the nineteenth century, including most of the geometers whose names are mentioned in the present article.*

Möbius began his contributions to geometry in 1823, and four years later published his Barycentrische Calcül. In this great work he introduced homogeneous coordinates with the

* For details see Loria, Il passato e il presente, etc.

attendant symmetry of geometric formulas, the scientific exposition of the principle of signs in geometry, and the establishment of the principle of geometric correspondence simple and multiple. He also (1852) summed up the classification of cubic curves, a service rendered by Zeuthen (1874) for quartics. To the period of Möbius also belong Bobillier (1827), who first used trilinear coordinates, and Bellavitis, whose contributions to analytic geometry were extensive. Gergonne's labors are mentioned in the next article.

Of all modern contributors to analytic geometry, Plücker stands foremost. In 1828 he published the first volume of his Analytisch-geometrische Entwickelungen, in which appeared the modern abridged notation, and which marks the beginning of a new era for analytic geometry. In the second volume (1831) he sets forth the present analytic form of the principle of duality. To him is due (1833) the general treatment of foci for curves of higher degree, and the complete classification of plane cubic curves (1835) which had been so frequently tried before him. He also gave (1839) an enumeration of plane curves of the fourth order, which Bragelogne and Euler had attempted. In 1842 he gave his celebrated "six equations" by which he showed that the characteristics of a curve (order, class, number of double points, number of cusps, number of double tangents, and number of inflections) are known when any three are given. To him is also due the first scientific dual definition of a curve, a system of tangential coordinates, and an investigation of the question of double tangents, a question further elaborated by Cayley (1847, 1858), Hesse (1847), Salmon (1858), and Dersch (1874). The theory of ruled surfaces, opened by Monge, was also extended by him. Possibly the greatest service rendered by Plücker was the introduction of the straight line as a space element, his first contribution (1865) being followed by his well-known treatise on the subject (1868–69). In this work he treats certain general properties of complexes, congruences, and ruled surfaces, as well as special properties of linear complexes and congruences, subjects also considered by Kummer and by Klein and others of the modern school. It is not a little due to Plücker that the concept of 4- and hence *n*-dimensional space, already suggested by Lagrange and Gauss, became the subject of later research. Riemann, Helmholtz, Lipschitz, Kronecker, Klein, Lie, Veronese, Cayley, d'Ovidio, and many others have elaborated the theory. The regular hypersolids in 4-dimensional space have been the subject of special study by Scheffler, Rudel, Hoppe, Schlegel, and Stringham.

Among Jacobi's contributions is the consideration (1836) of curves and groups of points resulting from the intersection of algebraic surfaces, a subject carried forward by Reye (1869). To Jacobi is also due the conformal representation of the ellipsoid on a plane, a treatment completed by Schering (1858). The number of examples of conformal representation of surfaces on planes or on spheres has been increased by Schwarz (1869) and Amstein (1872).

In 1844 Hesse, whose contributions to geometry in general are both numerous and valuable, gave the complete theory of inflections of a curve, and introduced the so-called Hessian curve as the first instance of a covariant of a ternary form. He also contributed to the theory of curves of the third order, and generalized the Pascal and Brianchon theorems on a spherical surface. Hesse's methods have recently been elaborated by Gundelfinger (1894).

Besides contributing extensively to synthetic geometry, Chasles added to the theory of curves of the third and fourth degrees. In the method of characteristics which he worked out may be found the first trace of the Abzählende Geometrie* which has been developed by Jonquières, Halphen (1875), and Schubert (1876, 1879), and to which Clebsch, Lindemann, and Hurwitz have also contributed. The general theory of correspondence starts with Geometry, and Chasles (1864) undertook

* Loria, G., Notizie storiche sulla Geometria numerativa. Bibliotheca Mathematica, 1888, pp. 39, 67; 1889, p. 23.

the first special researches on the correspondence of algebraic curves, limiting his investigations, however, to curves of deficiency zero. Cayley (1866) carried this theory to curves of higher deficiency, and Brill (from 1873) completed the theory.

Cayley's * influence on geometry was very great. He early carried on Plücker's consideration of singularities of a curve, and showed (1864, 1866) that every singularity may be considered as compounded of ordinary singularities so that the "six equations" apply to a curve with any singularities whatsoever. He thus opened a field for the later investigations of Noether, Zeuthen, Halphen, and H. J. S. Smith. Cayley's theorems on the intersection of curves (1843) and the determination of self-corresponding points for algebraic correspondences of a simple kind are fundamental in the present theory, subjects to which Bacharach, Brill, and Noether have also contributed extensively. Cayley added much to the theories of rational transformation and correspondence, showing the distinction between the theory of transformation of spaces and that of correspondence of loci. His investigations on the bitangents of plane curves, and in particular on the twenty-eight bitangents of a non-singular quartic, his developments of Plücker's conception of foci, his discussion of the osculating conics of curves and of the sextactic points on a plane curve, the geometric theory of the invariants and covariants of plane curves, are all noteworthy. He was the first to announce (1849) the twenty-seven lines which lie on a cubic surface, he extended Salmon's theory of reciprocal surfaces, and treated (1869) the classification of cubic surfaces, a subject already discussed by Schläfli. He also contributed to the theory of scrolls (skew-ruled surfaces), orthogonal systems of surfaces, the wave surface, etc., and was the first to reach (1845) any very general results in the theory of curves of double curvature, a theory in which the next great advance was made (1882) by Halphen and Noether. Among Cayley's other contributions to geometry is his theory of the Absolute, a figure in connection with which all metrical properties of a figure are considered.

* Biographical Notice in Cayley's Collected papers, Vol. VIII.

Clebsch * was also prominent in the study of curves and surfaces. He first applied the algebra of linear transformation to geometry. He emphasized the idea of deficiency (Geschlecht) of a curve, a notion which dates back to Abel, and applied the theory of elliptic and Abelian functions to geometry, using it for the study of curves. Clebsch (1872) investigated the shapes of surfaces of the third order. Following him, Klein attacked the problem of determining all possible forms of such surfaces, and established the fact that by the principle of continuity all forms of real surfaces of the third order can be derived from the particular surface having four real conical points. Zeuthen (1874) has discussed the various forms of plane curves of the fourth order, showing the relation between his results and those of Klein on cubic surfaces. Attempts have been made to extend the subject to curves of the *n*th order, but no general classification has been made. Quartic surfaces have been studied by Rohn (1887) but without a complete enumeration, and the same writer has contributed (1881) to the theory of Kummer surfaces.

Lie has adopted Plucker's generalized space element and extended the theory. His sphere geometry treats the subject from the higher standpoint of six homogeneous coordinates, as distinguished from the elementary sphere geometry with but five and characterized by the conformal group, a geometry studied by Darboux. Lie's theory of contact transformations, with its application to differential equations, his line and sphere complexes, and his work on minimum surfaces are all prominent.

Of great help in the study of curves and surfaces and of the theory of functions are the models prepared by Dyck, Brill, O. Henrici, Schwarz, Klein, Schönflies, Kummer, and others.†

The Theory of Minimum Surfaces has been developed along

* Klein, Evanston Lectures, Lect. I.

† Dyck, W., Katalog mathematischer und mathematisch-physikalischer Modelle, München, 1892; Deutsche Universitätsausstellung, Mathematical Papers of Chicago Congress, p. 49.

with the analytic geometry in general. Lagrange (1760–61) gave the equation of the minimum surface through a given contour, and Meusnier (1776, published in 1785) also studied the question. But from this time on for half a century little that is noteworthy was done, save by Poisson (1813) as to certain imaginary surfaces. Monge (1784) and Legendre (1787) connected the study of surfaces with that of differential equations, but this did not immediately affect this question. Scherk (1835) added a number of important results, and first applied the labors of Monge and Legendre to the theory. Catalan (1842), Björling (1844), and Dini (1865) have added to the subject. But the most prominent contributors have been Bonnet, Schwarz, Darboux, and Weierstrass. Bonnet (from 1853) has set forth a new system of formulas relative to the general theory of surfaces, and completely solved the problem of determining the minimum surface through any curve and admitting in each point of this curve a given tangent plane. Weierstrass (1866) has contributed several fundamental theorems, has shown how to find all of the real algebraic minimum surfaces, and has shown the connection between the theory of functions of an imaginary variable and the theory of minimum surfaces.

ART. 16. MODERN GEOMETRY.

Descriptive,* Projective, and Modern Synthetic Geometry are so interwoven in their historic development that it is even more difficult to separate them from one another than from the analytic geometry just mentioned. Monge had been in possession of his theory for over thirty years before the publication of his Géométrie Descriptive (1800), a delay due to the jealous desire of the military authorities to keep the valuable secret. It is true that certain of its features can be traced back to Desargues, Taylor, Lambert, and Frézier, but it was Monge who worked it out in detail as a science, although

* Wiener, Chr., Lehrbuch der darstellenden Geometrie, Leipzig, 1884–87; Geschichte der darstellenden Geometrie, 1884.

Lacroix (1795), inspired by Monge's lectures in the École Polytechnique, published the first work on the subject. After Monge's work appeared, Hachette (1812, 1818, 1821) added materially to its symmetry, subsequent French contributors being Leroy (1842), Olivier (from 1845), de la Gournerie (from 1860), Vallée, de Fourcy, Adhémar, and others. In Germany leading contributors have been Ziegler (1843), Anger (1858), and especially Fiedler (3d edn. 1883–88) and Wiener (1884–87). At this period Monge by no means confined himself to the descriptive geometry. So marked were his labors in the analytic geometry that he has been called the father of the modern theory. He also set forth the fundamental theorem of reciprocal polars, though not in modern language, gave some treatment of ruled surfaces, and extended the theory of polars to quadrics.*

Monge and his school concerned themselves especially with the relations of form, and particularly with those of surfaces and curves in a space of three dimensions. Inspired by the general activity of the period, but following rather the steps of Desargues and Pascal, Carnot treated chiefly the metrical relations of figures. In particular he investigated these relations as connected with the theory of transversals, a theory whose fundamental property of a four-rayed pencil goes back to Pappos, and which, though revived by Desargues, was set forth for the first time in its general form in Carnot's Géométrie de Position (1803), and supplemented in his Théorie des Transversales (1806). In these works he introduced negative magnitudes, the general quadrilateral and quadrangle, and numerous other generalizations of value to the elementary geometry of to-day. But although Carnot's work was important and many details are now commonplace, neither the name of the theory nor the method employed have endured. The present Geometry of Position (Geometrie der Lage) has little in common with Carnot's Géométrie de Position.

* On recent development of graphic methods and the influence of Monge upon this branch of mathematics, see Eddy, H. T., Modern Graphical Developments, Mathematical Papers of Chicago Congress (New York, 1896), p 58.

Projective Geometry had its origin somewhat later than the period of Monge and Carnot. Newton had discovered that all curves of the third order can be derived by central projection from five fundamental types. But in spite of this fact the theory attracted so little attention for over a century that its origin is generally ascribed to Poncelet. A prisoner in the Russian campaign, confined at Saratoff on the Volga (1812–14), "privé," as he says, "de toute espèce de livres et de secours, surtout distrait par les malheurs de ma patrie et les miens propres," he still had the vigor of spirit and the leisure to conceive the great work which he published (1822) eight years later. In this work was first made prominent the power of central projection in demonstration and the power of the principle of continuity in research. His leading idea was the study of projective properties, and as a foundation principle he introduced the anharmonic ratio, a concept, however, which dates back to Pappos and which Desargues (1639) had also used. Möbius, following Poncelet, made much use of the anharmonic ratio in his Barycentrische Calcül (1827), but under the name "Doppelschnitt-Verhältniss" (ratio bisectionalis), a term now in common use under Steiner's abbreviated form "Doppelverhältniss." The name "anharmonic ratio" or "function" (rapport anharmonique, or fonction anharmonique) is due to Chasles, and "cross-ratio" was coined by Clifford. The anharmonic point and line properties of conics have been further elaborated by Brianchon, Chasles, Steiner, and von Staudt. To Poncelet is also due the theory of "figures homologiques," the perspective axis and perspective center (called by Chasles the axis and center of homology), an extension of Carnot's theory of transversals, and the "cordes idéales" of conics which Plücker applied to curves of all orders. He also discovered what Salmon has called "the circular points at infinity," thus completing and establishing the first great principle of modern geometry, the principle of continuity. Brianchon (1806), through his application of Desargues's theory of polars, completed the foundation which Monge had begun for Poncelet's (1829) theory of reciprocal polars.

Among the most prominent geometers contemporary with Poncelet was Gergonne, who with more propriety might be ranked as an analytic geometer. He first (1813) used the term "polar" in its modern geometric sense, although Servois (1811) had used the expression "pole." He was also the first (1825–26) to grasp the idea that the parallelism which Maurolycus, Snell, and Viète had noticed is a fundamental principle. This principle he stated and to it he gave the name which it now bears, the Principle of Duality, the most important, after that of continuity, in modern geometry. This principle of geometric reciprocation, the discovery of which was also claimed by Poncelet, has been greatly elaborated and has found its way into modern algebra and elementary geometry, and has recently been extended to mechanics by Genese. Gergonne was the first to use the word "class" in describing a curve, explicitly defining class and degree (order) and showing the duality between the two. He and Chasles were among the first to study scientifically surfaces of higher order.

Steiner (1832) gave the first complete discussion of the projective relations between rows, pencils, etc., and laid the foundation for the subsequent development of pure geometry. He practically closed the theory of conic sections, of the corresponding figures in three-dimensional space and of surfaces of the second order, and hence with him opens the period of special study of curves and surfaces of higher order. His treatment of duality and his application of the theory of projective pencils to the generation of conics are masterpieces. The theory of polars of a point in regard to a curve had been studied by Bobillier and by Grassmann, but Steiner (1848) showed that this theory can serve as the foundation for the study of plane curves independently of the use of coordinates, and introduced those noteworthy curves covariant to a given curve which now bear the names of himself, Hesse, and Cayley. This whole subject has been extended by Grassmann, Chasles,

Cremona, and Jonquières. Steiner was the first to make prominent (1832) an example of correspondence of a more complicated nature than that of Poncelet, Möbius, Magnus, and Chasles. His contributions, and those of Gudermann, to the geometry of the sphere were also noteworthy.

While Möbius, Plücker, and Steiner were at work in Germany, Chasles was closing the geometric era opened in France by Monge. His Aperçu Historique (1837) is a classic, and did for France what Salmon's works did for algebra and geometry in England, popularizing the researches of earlier writers and contributing both to the theory and the nomenclature of the subject. To him is due the name "homographic" and the complete exposition of the principle as applied to plane and solid figures, a subject which has received attention in England at the hands of Salmon, Townsend, and H. J. S. Smith.

Von Staudt began his labors after Plücker, Steiner, and Chasles had made their greatest contributions, but in spite of this seeming disadvantage he surpassed them all. Joining the Steiner school, as opposed to that of Plücker, he became the greatest exponent of pure synthetic geometry of modern times. He set forth (1847, 1856–60) a complete, pure geometric system in which metrical geometry finds no place. Projective properties foreign to measurements are established independently of number relations, number being drawn from geometry instead of conversely, and imaginary elements being systematically introduced from the geometric side. A projective geometry based on the group containing all the real projective and dualistic transformations, is developed, imaginary transformations being also introduced. Largely through his influence pure geometry again became a fruitful field. Since his time the distinction between the metrical and projective theories has been to a great extent obliterated,* the metrical properties

* Klein, F., Erlangen Programme of 1872, Haskell's translation, Bulletin of New York Mathematical Society, Vol. II, p. 215.

being considered as projective relations to a fundamental configuration, the circle at infinity common for all spheres. Unfortunately von Staudt wrote in an unattractive style, and to Reye is due much of the popularity which now attends the subject.

Cremona began his publications in 1862. His elementary work on projective geometry (1875) in Leudesdorf's translation is familiar to English readers. His contributions to the theory of geometric transformations are valuable, as also his works on plane curves, surfaces, etc.

In England Mulcahy, but especially Townsend (1863), and Hirst, a pupil of Steiner's, opened the subject of modern geometry. Clifford did much to make known the German theories, besides himself contributing to the study of polars and the general theory of curves.

ART. 17. ELEMENTARY GEOMETRY.

Trigonometry and Elementary Geometry have also been affected by the general mathematical spirit of the century. In trigonometry the general substitution of ratios for lines in the definitions of functions has simplified the treatment, and certain formulas have been improved and others added.* The convergence of trigonometric series, the introduction of the Fourier series, and the free use of the imaginary have already been mentioned. The definition of the sine and cosine by series, and the systematic development of the theory on this basis, have been set forth by Cauchy (1821), Lobachevsky (1833), and others. The hyperbolic trigonometry,† already founded by Mayer and Lambert, has been popularized and further developed by Gudermann (1830), Hoüel, and Laisant (1871), and projective formulas and generalized figures have

* Todhunter, I., History of certain formulas of spherical trigonometry, Philosophical Magazine, 1873.

† Günther, S., Die Lehre von den gewöhnlichen und verallgemeinerten Hyperbelfunktionen, Halle, 1881; Chrystal, G., Algebra, Vol. II, p. 288.

been introduced, notably by Gudermann, Möbius, Poncelet, and Steiner. Recently Study has investigated the formulas of spherical trigonometry from the standpoint of the modern theory of functions and theory of groups, and Macfarlane has generalized the fundamental theorem of trigonometry for three-dimensional space.

Elementary Geometry has been even more affected. Among the many contributions to the theory may be mentioned the following: That of Möbius on the opposite senses of lines, angles, surfaces, and solids; the principle of duality as given by Gergonne and Poncelet; the contributions of De Morgan to the logic of the subject; the theory of transversals as worked out by Monge, Brianchon, Servois, Carnot, Chasles, and others; the theory of the radical axis, a property discovered by the Arabs, but introduced as a definite concept by Gaultier (1813) and used by Steiner under the name of "line of equal power"; the researches of Gauss concerning inscriptible polygons, adding the 17- and 257-gon to the list below the 1000-gon; the theory of stellar polyhedra as worked out by Cauchy, Jacobi, Bertrand, Cayley, Möbius, Wiener, Hess, Hersel, and others, so that a whole series of bodies have been added to the four Kepler-Poinsot regular solids; and the researches of Muir on stellar polygons. These and many other improvements now find more or less place in the text-books of the day.

To these must be added the recent Geometry of the Triangle, now a prominent chapter in elementary mathematics. Crelle (1816) made some investigations in this line, Feuerbach (1822) soon after discovered the properties of the Nine-Point Circle, and Steiner also came across some of the properties of the triangle, but none of these followed up the investigation. Lemoine * (1873) was the first to take up the subject in a sys-

* Smith, D. E., Biography of Lemoine, American Mathematical Monthly, Vol. III, p. 29; Mackay, J. S., various articles on modern geometry in Proceedings Edinburgh Mathematical Society, various years; Vigarié, É., Géométrie du triangle. Articles in recent numbers of Journal de Mathématiques spéciales, Mathesis, and Proceedings of the Association française pour l'avancement des sciences.

tematic way, and he has contributed extensively to its development. His theory of "transformation continue" and his "géométrographie" should also be mentioned. Brocard's contributions to the geometry of the triangle began in 1877. Other prominent writers have been Tucker, Neuberg, Vigarié, Emmerich, M'Cay, Longchamps, and H. M. Taylor. The theory is also greatly indebted to Miller's work in The Educational Times, and to Hoffmann's Zeitschrift.

The study of linkages was opened by Peaucellier (1864), who gave the first theoretically exact method for drawing a straight line. Kempe and Sylvester have elaborated the subject.

In recent years the ancient problems of trisecting an angle, doubling the cube, and squaring the circle have all been settled by the proof of their insolubility through the use of compasses and straight edge.*

ART. 18. NON-EUCLIDEAN GEOMETRY.

The Non-Euclidean Geometry† is a natural result of the futile attempts which had been made from the time of Proklos to the opening of the nineteenth century to prove the fifth postulate (also called the twelfth axiom, and sometimes the

* Klein, F., Vorträge über ausgewählten Fragen; Rudio, F., Das Problem von der Quadratur des Zirkels. Naturforschende Gesellschaft Vierteljahrschrift, 1890; Archimedes, Huygens, Lambert, Legendre (Leipzig, 1892).

† Stäckel and Engel, Die Theorie der Parallellinien von Euklid bis auf Gauss, Leipzig, 1895; Halsted, G. B., various contributions: Bibliography of Hyperspace and Non-Euclidean Geometry, American Journal of Mathematics, Vols. I, II; The American Mathematical Monthly, Vol. I; translations of Lobachevsky's Geometry, Vasiliev's address on Lobachevsky, Saccheri's Geometry, Bolyai's work and his life; Non-Euclidean and Hyperspaces, Mathematical Papers of Chicago Congress, p. 92. Loria, G., Die hauptsächlichsten Theorien der Geometrie, p. 106; Karagiannides, A., Die Nichteuklidische Geometrie vom Alterthum bis zur Gegenwart, Berlin, 1893; McClintock, E., On the early history of Non-Euclidean Geometry, Bulletin of New York Mathematical Society, Vol. II, p. 144; Poincaré, Non-Euclidean Geom., Nature, 45:404; Articles on Parallels and Measurement in Encyclopædia Britannica, 9th edition; Vasiliev's address (German by Engel) also appears in the Abhandlungen zur Geschichte der Mathematik, 1895.

eleventh or thirteenth) of Euclid. The first scientific investigation of this part of the foundation of geometry was made by Saccheri (1733), a work which was not looked upon as a precursor of Lobachevsky, however, until Beltrami (1889) called attention to the fact. Lambert was the next to question the validity of Euclid's postulate, in his Theorie der Parallellinien (posthumous, 1786), the most important of many treatises on the subject between the publication of Saccheri's work and those of Lobachevsky and Bolyai. Legendre also worked in the field, but failed to bring himself to view the matter outside the Euclidean limitations.

During the closing years of the eighteenth century Kant's* doctrine of absolute space, and his assertion of the necessary postulates of geometry, were the object of much scrutiny and attack. At the same time Gauss was giving attention to the fifth postulate, though on the side of proving it. It was at one time surmised that Gauss was the real founder of the non-Euclidean geometry, his influence being exerted on Lobachevsky through his friend Bartels, and on Johann Bolyai through the father Wolfgang, who was a fellow student of Gauss's. But it is now certain that Gauss can lay no claim to priority of discovery, although the influence of himself and of Kant, in a general way, must have had its effect.

Bartels went to Kasan in 1807, and Lobachevsky was his pupil. The latter's lecture notes show that Bartels never mentioned the subject of the fifth postulate to him, so that his investigations, begun even before 1823, were made on his own motion and his results were wholly original. Early in 1826 he sent forth the principles of his famous doctrine of parallels, based on the assumption that through a given point more than one line can be drawn which shall never meet a given line coplanar with it. The theory was published in full in 1829–30, and he contributed to the subject, as well as to other branches of mathematics, until his death.

* Fink, E., Kant als Mathematiker, Leipzig, 1889.

Johann Bolyai received through his father, Wolfgang, some of the inspiration to original research which the latter had received from Gauss. When only twenty-one he discovered, at about the same time as Lobachevsky, the principles of non-Euclidean geometry, and refers to them in a letter of November, 1823. They were committed to writing in 1825 and published in 1832. Gauss asserts in his correspondence with Schumacher (1831–32) that he had brought out a theory along the same lines as Lobachevsky and Bolyai, but the publication of their works seems to have put an end to his investigations. Schweikart was also an independent discoverer of the non-Euclidean geometry, as his recently recovered letters show, but he never published anything on the subject, his work on the theory of parallels (1807), like that of his nephew Taurinus (1825), showing no trace of the Lobachevsky-Bolyai idea.

The hypothesis was slowly accepted by the mathematical world. Indeed it was about forty years after its publication that it began to attract any considerable attention. Hoüel (1866) and Flye St. Marie (1871) in France, Riemann (1868), Helmholtz (1868), Frischauf (1872), and Baltzer (1877) in Germany, Beltrami (1872) in Italy, de Tilly (1879) in Belgium, Clifford in England, and Halsted (1878) in America, have been among the most active in making the subject popular. Since 1880 the theory may be said to have become generally understood and accepted as legitimate.*

Of all these contributions the most noteworthy from the scientific standpoint is that of Riemann. In his Habilitationsschrift (1854) he applied the methods of analytic geometry to the theory, and suggested a surface of negative curvature, which Beltrami calls "pseudo-spherical," thus leaving Euclid's geometry on a surface of zero curvature midway between his own and Lobachevsky's. He thus set forth three kinds of

* For an excellent summary of the results of the hypothesis, see an article by McClintock, The Non-Euclidian Gecmetry, Bulletin of New York Mathematical Society, Vol. II, p. 1.

geometry, Bolyai having noted only two. These Klein (1871) has called the elliptic (Riemann's), parabolic (Euclid's), and hyperbolic (Lobachevsky's).

Starting from this broader point of view* there have contributed to the subject many of the leading mathematicians of the last quarter of a century, including, besides those already named, Cayley, Lie, Klein, Newcomb, Pasch, C. S. Peirce, Killing, Fiedler, Mansion, and McClintock. Cayley's contribution of his "metrical geometry" was not at once seen to be identical with that of Lobachevsky and Bolyai. It remained for Klein (1871) to show this, thus simplifying Cayley's treatment and adding one of the most important results of the entire theory. Cayley's metrical formulas are, when the Absolute is real, identical with those of the hyperbolic geometry; when it is imaginary, with the elliptic; the limiting case between the two gives the parabolic (Euclidean) geometry. The question raised by Cayley's memoir as to how far projective geometry can be defined in terms of space without the introduction of distance had already been discussed by von Staudt (1857) and has since been treated by Klein (1873) and by Lindemann (1876).

ART. 19. BIBLIOGRAPHY.

The following are a few of the general works on the history of mathematics in the nineteenth century, not already mentioned in the foot-notes. For a complete bibliography of recent works the reader should consult the Jahrbuch über die Fortschritte der Mathematik, the Bibliotheca Mathematica, or the Revue Semestrielle, mentioned below.

Abhandlungen zur Geschichte der Mathematik (Leipzig).

Ball, W. W. R., A short account of the history of mathematics (London, 1893).

Ball, W. W. R., History of the study of mathematics at Cambridge (London, 1889).

Ball, W. W. R., Primer of the history of mathematics (London, 1895).

* Klein, Evanston Lectures, Lect. IX.

Bibliotheca Mathematica, G. Eneström, Stockholm. Quarterly. Should be consulted for bibliography of current articles and works on history of mathematics.

Bulletin des Sciences Mathématiques (Paris, IIième Partie).

Cajori, F., History of Mathematics (New York, 1894).

Cayley, A., Inaugural address before the British Association, 1883. Nature, Vol. XXVIII, p. 491.

Dictionary of National Biography. London, not completed. Valuable on biographies of British mathematicians.

D'Ovidio, Enrico, Uno sguardo alle origini ed allo sviluppo della Matematica Pura (Torino, 1889).

Dupin, Ch., Coup d'œil sur quelques progrès des Sciences mathématiques, en France, 1830–35. Comptes Rendus, 1835.

Encyclopædia Britannica. Valuable biographical articles by Cayley, Chrystal, Clerke, and others.

Fink, K., Geschichte der Mathematik (Tübingen, 1890). Bibliography on p. 255.

Gerhardt, C. J., Geschichte der Mathematik in Deutschland (Munich, 1877).

Graf, J. H., Geschichte der Mathematik und der Naturwissenschaften in bernischen Landen (Bern, 1890). Also numerous biographical articles.

Günther, S., Vermischte Untersuchungen zur Geschichte der mathematischen Wissenschaften (Leipzig, 1876).

Günther, S., Ziele und Resultate der neueren mathematisch-historischen Forschung (Erlangen, 1876).

Hagen, J. G., Synopsis der höheren Mathematik. Two volumes (Berlin, 1891–93).

Hankel, H., Die Entwickelung der Mathematik in dem letzten Jahrhundert (Tübingen, 1884).

Hermite, Ch., Discours prononcé devant le président de la république le 5 août 1889 à l'inauguration de la nouvelle Sorbonne. Bulletin des Sciences mathématiques, 1890 ; also Nature, Vol. XLI, p. 597. (History of nineteenth-century mathematics in France.)

Hoefer, F., Histoire des mathématiques (Paris, 1879).

Isely, L., Essai sur l'histoire des mathématiques dans la Suisse française (Neuchâtel, 1884).

Jahrbuch über die Fortschritte der Mathematik (Berlin, annually, 1868 to date).

Marie, M., Histoire des sciences mathématiques et physiques. Vols. X, XI, XII (Paris, 1887–88).

Matthiessen, L., Grundzüge der antiken und modernen Algebra der litteralen Gleichungen (Leipzig, 1878).

Newcomb, S., Modern mathematical thought. Bulletin New York Mathematical Society, Vol. III, p. 95; Nature, Vol. XLIX, p. 325.

Poggendorff, J. C., Biographisch-literarisches Handwörterbuch zur Geschichte der exacten Wissenschaften. Two volumes (Leipzig, 1863), and two later supplementary volumes.

Quetelet, A., Sciences mathématiques et physiques chez les Belges au commencement du XIXe siècle (Brussels, 1866).

Revue semestrielle des publications mathématiques rédigée sous les auspices de la Société mathématique d'Amsterdam. 1893 to date. (Current periodical literature.)

Roberts, R. A., Modern mathematics. Proceedings of the Irish Academy, 1888.

Smith, H. J. S., On the present state and prospects of some branches of pure mathematics. Proceedings of London Mathematical Society, 1876; Nature, Vol. XV, p. 79.

Sylvester, J. J., Address before the British Association. Nature, Vol. I, pp. 237, 261.

Wolf, R., Handbuch der Mathematik. Two volumes (Zurich, 1872).

Zeitschrift für Mathematik und Physik. Historisch-literarische Abtheilung. Leipzig. The Abhandlungen zur Geschichte der Mathematik are supplements.

For a biographical table of mathematicians see Fink's Geschichte der Mathematik, p. 240. For the names and positions of living mathematicians see the Jahrbuch der gelehrten Welt, published at Strassburg.

Since the above bibliography was prepared the nineteenth century has closed. With its termination there would naturally be expected a series of retrospective views of the development of a hundred years in all lines of human progress. This expectation was duly fulfilled, and numerous addresses and memoirs testify to the interest recently awakened in the subject. Among the contributions to the general history of modern mathematics may be cited the following:

Pierpont, J., St. Louis address, 1904. Bulletin of the American Mathematical Society (N. S.), Vol. IX, p. 136. An excellent survey of the century's progress in pure mathematics.

Günther, S., Die Mathematik im neunzehnten Jahrhundert. Hoffmann's Zeitschrift, Vol. XXXII, p. 227.

Adhémar, R. d', L'œuvre mathématique du XIXe siècle. Revue des questions scientifiques, Louvain Vol. XX (2), p. 177 (1901).

Picard, E., Sur le développement, depuis un siècle, de quelques théories fondamentales dans l'analyse mathématique. Conférences faite à Clark University (Paris, 1900).

Lampe, E., Die reine Mathematik in den Jahren 1884–1899 (Berlin, 1900).

Among the contributions to the history of applied mathematics in general may be mentioned the following:

Woodward, R. S., Presidential address before the American Mathematical Society in December, 1899. Bulletin of the American Mathematical Society (N. S.), Vol. VI, p. 133. (German, in the Naturwiss. Rundschau, Vol. XV; Polish, in the Wiadomości Matematyczne, Warsaw, Vol. V (1901).) This considers the century's progress in applied mathematics.

Mangoldt, H. von, Bilder aus der Entwickelung der reinen und angewandten Mathematik während des neunzehnten Jahrhunderts mit besonderer Berücksichtigung des Einflusses von Carl Friedrich Gauss. Festrede (Aachen, 1900).

Van t' Hoff, J. H., Ueber die Entwickelung der exakten Naturwissenschaften im 19. Jahrhundert. Vortrag gehalten in Aachen, 1900. Naturwiss. Rundschau, Vol. XV, p. 557 (1900).

The following should be mentioned as among the latest contributions to the history of modern mathematics in particular countries:

Fiske, T. S., Presidential address before the American Mathematical Society in December, 1904. Bulletin of the American Mathematical Society (N. S.), Vol. IX, p. 238. This traces the development of mathematics in the United States.

Purser, J., The Irish school of mathematicians and physicists from the beginning of the nineteenth century. Nature, Vol. LXVI, p. 478 (1902).

Guimarães, R. Les mathématiques en Portugal au XIXe siècle. (Coïmbre, 1900).

A large number of articles upon the history of special branches of mathematics have recently appeared, not to mention the custom of inserting historical notes in the recent treatises upon the subjects themselves. Of the contributions to the history of particular branches, the following may be mentioned as types:

Miller, G. A., Reports on the progress in the theory of groups of a finite order. Bulletin of the American Mathematical Society (N. S.), Vol. V, p. 227; Vol. IX, p. 106. Supplemental report by Dickson, L. E., Vol. VI, p. 13, whose treatise on Linear Groups (1901) is a history in itself. Steinitz and Easton have also contributed to this subject.

Hancock, H., On the historical development of the Abelian functions to the time of Riemann. British Association Report for 1897.

Brocard, H., Notes de bibliographie des courbes géométriques. Bar-le-Duc, 2 vols., lithog., 1897, 1899.

Hagen, J. G., On the history of the extensions of the calculus. Bulletin of the American Mathematical Society (N. S.), Vol. VI, p. 381.

Hill, J. E., Bibliography of surfaces and twisted curves. Ib., Vol. III, p. 133 (1897).

Aubry, A., Historia del problema de las tangentes. El Progresso matematico, Vol. I (2), pp. 129, 164.

Compère, C., Le problème des brachistochrones. Essai historique. Mémoires de la Société d. Sciences, Liège, Vol. I (3), p. 128 (1899).

Stäckel, P., Beiträge zur Geschichte der Funktionentheorie im achtzehnten Jahrhundert. Bibliotheca Mathematica, Vol. II (3), p. 111 (1901).

Obenrauch, F. J., Geschichte der darstellenden und projektiven Geometrie mit besonderer Berücksichtigung ihrer Begründung in Frankreich und Deutschland und ihrer wissenschaftlichen Pflege in Oesterreich (Brünn, 1897).

Muir, Th., The theory of alternants in the historical order of its development up to 1841. Proceedings of the Royal Society of Edinburgh, Vol. XXIII (2), p. 93 (1899). The theory of screw determinants and Pfaffians in the historical order of its development up to 1857. Ib., p. 181.

Papperwitz, E., Ueber die wissenschaftliche Bedeutung der darstellenden Geometrie und ihre Entwickelung bis zur systematischen Begründung durch Gaspard Monge. Rede (Freiberg i./S., 1901).

Mention should also be made of the fact that the Bibliotheca Mathematica, a journal devoted to the history of the mathematical sciences, began its third series in 1900. It remains under the able editorship of G. Eneström, and in its new series it appears in much enlarged form. It contains numerous articles on the history of modern mathematics, with a complete current bibliography of this field.

Besides direct contributions to the history of the subject, and historical and bibliographical notes, several important works have recently appeared which are historical in the best sense, although written from the mathematical standpoint. Of these there are three that deserve special mention:

Encyklopädie der mathematischen Wissenschaften mit Einschluss ihrer Anwendungen. The publication of this monumental work was begun in 1898, and the several volumes are being carried on simultaneously. The first volume (Arithmetik and Algebra) was completed in 1904. This publication is under the patronage of the academies of sciences of Göttingen, Leipzig, Munich, and Vienna. A French translation, with numerous additions, is in progress.

Pascal, E., Repertorium der höheren Mathematik, translated from the Italian by A. Schepp. Two volumes (Leipzig, 1900, 1902). It contains an excellent bibliography, and is itself a history of modern mathematics.

Hagen, J. G., Synopsis der höheren Mathematik. This has been for some years in course of publication, and has now completed Vol. III.

In the line of biography of mathematicians, with lists of published works, Poggendorff's Biographisch-literarisches Handwörterbuch zur Geschichte der exacten Wissenschaften has reached its fourth volume (Leipzig, 1903), this volume covering the period from 1883 to 1902. A new biographical table has been added to the English translation of Fink's History of Mathematics (Chicago, 1900).

Art. 20. General Tendencies.

The opening of the nineteenth century was, as we have seen, a period of profound introspection following a period of somewhat careless use of the material accumulated in the seventeenth century. The mathematical world sought to orientate itself, to examine the foundations of its knowledge, and to critically examine every step in its several theories. It then took up the line of discovery once more, less recklessly than before, but still with thoughts directed primarily in the direction of invention. At the close of the century there came again a period of introspection, and one of the recent tendencies is towards a renewed study of foundation principles. In England one of the leaders in this movement is Russell, who has studied the foundations of geometry (1897) and of mathematics in general (1903). In America the fundamental conceptions and methods of mathematics have been considered by Bôcher in his St. Louis address in 1904,* and the question of a series of irreducible postulates has been studied by Huntington. In Italy, Padoa and Bureli-Forti have studied the fundamental postulates of algebra, and Pieri those of geometry. In Germany, Hilbert has probably given the most attention to the foundation principles of geometry (1899), and more recently he has investigated the compatibility of the arithmetical axioms (1900). In France, Poincaré has considered the rôle of intuition and of logic in mathematics,† and in every country the foundation principles have been made the object of careful investigation.

As an instance of the orientation already mentioned, the noteworthy address of Hilbert at Paris in 1900 ‡ stands out prominently. This address reviews the field of pure mathematics and sets forth several of the greatest questions demanding investigation at the present time. In the particular line of geometry the

* Bulletin of the American Mathematical Society (N. S.), Vol. XI, p. 115.

† Compte rendu du deuxième congrès international des mathématiciens tenu à Paris, 1900. Paris, 1902, p. 115.

‡ Göttinger Nachrichten, 1900, p. 253; Archiv der Mathematik und Physik, 1901, pp. 44, 213; Bulletin of the American Mathematical Society, 1902, p. 437.

memoir which Segré wrote in 1891, on the tendencies in geometric investigation, has recently been revised and brought up to date.*

There is also seen at the present time, as never before, a tendency to coöperate, to exchange views, and to internationalize mathematics. The first international congress of mathematicians was held at Zurich in 1897, the second one at Paris in 1900, and the third at Heidelberg in 1904. The first international congress of philosophy was held at Paris in 1900, the third section being devoted to logic and the history of the sciences (on this occasion chiefly mathematics), and the second congress was held at Geneva in 1904. There was also held an international congress of historic sciences at Rome in 1903, an international committee on the organization of a congress on the history of sciences being at that time formed. The result of such gatherings has been an exchange of views in a manner never before possible, supplementing in an inspiring way the older form of international communication through published papers.

In the United States there has been shown a similar tendency to exchange opinions and to impart verbal information as to recent discoveries. The American Mathematical Society, founded in 1894,† has doubled its membership in the past decade,‡ and has increased its average of annual papers from 30 to 150. It has also established two sections, one at Chicago (1897) and one at San Francisco (1902). The activity of its members and the quality of papers prepared has led to the publication of the *Transactions*, beginning with 1900. In order that its members may be conversant with the lines of investigation in the various mathematical centers, the society publishes in its *Bulletin* the courses in advanced mathematics offered in many of the leading universities of the world. Partly as a result of this activity, and partly because of the large number of American students who have recently studied abroad, a remarkable change is at

* Bulletin of the American Mathematical Society (N. S.), Vol. X, p. 443.

† It was founded as the New York Mathematical Society six years earlier, in 1888.

‡ It is now, in 1905, approximately 500.

present passing over the mathematical work done in the universities and colleges of this country. Courses that a short time ago were offered in only a few of our leading universities are now not uncommon in institutions of college rank. They are often given by men who have taken advanced degrees in mathematics at Göttingen, Berlin, Paris, or other leading universities abroad, and they are awakening a great interest in the modern field. A recent investigation (1903) showed that 67 students in ten American institutions were taking courses in the theory of functions, 11 in the theory of elliptic functions, 94 in projective geometry, 26 in the theory of invariants, 45 in the theory of groups, and 46 in the modern advanced theory of equations, courses which only a few years ago were rarely given in this country. A similar change is seen in other countries, notably in England and Italy, where courses that a few years ago were offered only in Paris or in Germany are now within the reach of university students at home. The interest at present manifested by American scholars is illustrated by the fact that only four countries (Germany, Russia, Austria, and France) had more representatives than the United States, among the 336 regular members at the third international mathematical congress at Heidelberg in 1904.

The activity displayed at the present time in putting the work of the masters into usable form, so as to define clear points of departure along the several lines of research, is seen in the large number of collected works published or in course of publication in the last decade. These works have usually been published under governmental patronage, often by some learned society, and always under the editorship of some recognized authority. They include the works of Galileo, Fermat, Descartes, Huygens, Laplace, Gauss, Galois, Cauchy, Hesse, Plücker, Grassmann, Dirichlet, Laguerre, Kronecker, Fuchs, Weierstrass, Stokes, Tait, and various other leaders in mathematics. It is only natural to expect a number of other sets of collected works in the near future, for not only is there the remote past to draw upon, but the death roll of the last decade has been a large one. The following is only a partial list of eminent mathematicians who have recently died, and whose collected works have been or are in the course of being published, or may be deemed worthy of publication in the future: Cayley (1895), Neumann (1895), Tisserand (1896), Brioschi (1897), Sylvester (1897), Weierstrass (1897), Lie (1899), Beltrami (1900), Bertrand (1900), Tait (1901), Hermite (1901), Fuchs (1902), Gibbs (1903), Cremona (1903), and Salmon (1904), besides such writers as Frost (1898), Hoppe (1900), Craig (1900), Schlömilch (1901), Everett on the side of mathematical physics (1904), and Paul Tannery, the best of the modern French historians of mathematics (1904).*

It is, of course, impossible to detect with any certainty the present tendencies in mathematics. Judging, however, by the number and nature of the published papers and works of the past few years, it is reasonable to expect a great development in all lines, especially in such modern branches as the theory of groups, theory of functions, theory of invariants, higher geometry, and differential equations. If we may judge from the works in applied mathematics which have recently appeared, we are entering upon an era similar to that in which Laplace labored, an era in which all these modern theories of mathematics shall find application in the study of physical problems, including those that relate to the latest discoveries. The profound study of applied mathematics in France and England, the advanced work in discovery in pure mathematics in Germany and France, and the search for the logical bases for the science that has distinguished Italy as well as Germany, are all destined to affect the character of the international mathematics of the immediate future. Probably no single influence will be more prominent in the internationalizing process than the tendency of the younger generation of American mathematicians to study in England, France, Germany, and Italy, and to assimilate the best that each of these countries has to offer to the world.

* For students wishing to investigate the work of mathematicians who died in the last two decades of the nineteenth century, Eneström's "Bio-bibliographie der 1881–1900 verstorbenen Mathematiker," in the Bibliotheca Mathematica Vol. II (3), p. 326 (1901), will be found valuable.

INDEX.

Abelian functions, 45.
Abel's quintic, 20.
Absolute, the, 56.
Alternants, 27.
Analytic geometry, 52.
Anharmonic ratio, 60.
Approximation of roots, 19.
Ausdehnungslehre, 17.
Automorphic functions, 49.

Bibliography, 8, 68, 70.
Binary forms, 13, 31.
Binomial equations, 22.
Bessel's functions, 49.

Canonical forms, 31.
Calculus, 31.
  of variations, 33.
Complex numbers, 15.
Congruences, 12.
Conics, 61.
Continued fractions, 14.
Convergence, of series, 40.
  uniform, 41.
Covariants, 29.
Cubic surfaces, 56.
Curves, 52.

Deficiency curves, 48.
Definite integrals, 32.
De Moivre's formula, 15.
Descriptive geometry, 58.
Determinants, 26.
Differential calculus, 31.
  equations, 35.
Differential invariants, 31.
Discontinuous groups, 26.
Duality, principle of, 54.

Elementary geometry, 64.
Elimination, 23.
Elliptic functions, 22, 44.
  modular functions, 49.
Equations, binomial, 22.
  differential, 35.
  fundamental theorem, 20.
  modular, 25, 46.
  Plücker's, 56.
  quintic, 20, 25.
  theory of, 19.
Errors, theory of, 50.
Eulerian integrals, 32.

Forms, 28.
Fourier's series, 42.
Fractions, continued, 14.
Functions, Abelian, 45.
  Bernoulli's, 41.
  Bessel's, 49.
  canonical, 31.
  elliptic, 22, 44.
  elliptic modular, 49.
  gamma, 33.
  hyperelliptic, 46.
  Laplace's, 50.
  omega, 47.
  periodic, 47.
  symbols for, 43.
  symmetric, 23.

Galois's group theory, 25.
Gamma functions, 33.
General tendencies, 74.
Geometry, analytic, 52.
  descriptive, 58.
  elementary, 64.
  modern, 58.
  non-Euclidean, 65.
  projective, 60.
Groups, 24.
  of equations, 21.

Hessian curve, 55.
Histories of modern mathematics, 68.
Horner's method, 19.
Hyperdeterminants, 25.
Hyperelliptic functions, 46.
Hypergeometric series, 40.

Icosahedron equation, 21.
Imaginaries, 16.
Infinitesimal calculus, 34.
Infinite series, 39.
  convergence of, 40.
Integral calculus, 31.
  Abelian, 45.
International congresses, 75.
Interpolation formulas, 41.
Invariants, 29, 46.
Irrational numbers, 13.

Journals, mathematical, 9.

Lagrange's resolvent, 20.
  series, 49.
Laplace's functions, 50.
Law of reciprocity, 12.
Least squares, 50.

Mathematical bibliography, 68.
  periodicals, 9.
  works, 76.
Mathematicians, recent, 70, 77.
Mathematics, modern, 7.
Matrices, 19.
Metrical geometry, 68.
Minimum surfaces, 57.
Mittag-Leffler's theorem, 48.
Models, 57.
Modern geometry, 58.
  mathematics, 7.

Non-Euclidean geometry, 65.
Notations for functions, 43.
Numbers, complex, 15.
  prime, 12.
  theory of, 11, 13.
Numerical equations, 19.

Observations, errors of, 50.
Omega functions, 47.

Parallels, 66.
Partial differential equations, 35.
Periodicals, 9.
Periodic functions, 47.
Physics, 34.
Plücker's equations, 54, 56.
Pole and polar, 59, 61.
Polygons, 64.
Polyhedra, 64.
Potential, 34.
Prime numbers, 12.
Probabilities, 50.
Projective geometry, 60.

Quadratic forms, 13.
  residues, 12.
Quantics, 28.
Quantity, complex, 15.
Quaternions, 17.
Quintic equations, 20, 25.

Reciprocity, 12.
Removal of terms, 20.
Residues, 12.
Resolvents, 20.
Resultants, 23.
Roots of equations, 19.

Schools of mathematics, 9.
Semiconvergent series, 34.

Series, convergence of, 40.
Fourier's, 42.
hypergeometric, 40.
infinite, 39.
Lagrange's, 49.
semiconvergent, 34.
sine, 42.
trigonometric, 42.
Sextic resolvent, 20.
Sine series, 42.
Singular solutions, 37.
Solution of equations, 19.
Sturm's theorem, 19.
Substitutions, 24.
Surfaces, 53, 57.
cubic, 56.
minimum, 57.
Symbolic methods, 33.
Symmetric determinants, 28.
functions, 23.
Synthetic geometry, 52.

Ternary forms, 9.
Theory of functions, 43.
of numbers, 11.
Theta functions, 46.
Transcendent functions, 38.
numbers, 13.
Triangle, geometry of, 64.
Trigonometry, 59.
Trilinear coördinates, 54.

Uniform convergence, 41.
United States, outlook in the, 75.

Variations, calculus of, 33.
Vector analysis, 18.

Weierstrass's function, 48.
Works of mathematicians, collected, 76.

COSIMO

CPSIA information can be obtained at www.ICGtesting.com
Printed in the USA
BVOW04s1113280214

346311BV00007B/138/A